Anti-Disturbance Control for Systems with Multiple Disturbances

AUTOMATION AND CONTROL ENGINEERING
A Series of Reference Books and Textbooks

Series Editors

FRANK L. LEWIS, Ph.D.,
Fellow IEEE, Fellow IFAC
Professor
The Univeristy of Texas Research Institute
The University of Texas at Arlington

SHUZHI SAM GE, Ph.D.,
Fellow IEEE
Professor
Interactive Digital Media Institute
The National University of Singapore

PUBLISHED TITLES

Anti-Disturbance Control for Systems with Multiple Disturbances,
Lei Guo; Songyin Cao

Tensor Product Model Transformation in Polytopic Model-Based Control,
Péter Baranyi; Yeung Yam; Péter Várlaki

Fundamentals in Modeling and Control of Mobile Manipulators, *Zhijun Li; Shuzhi Sam Ge*

Optimal and Robust Scheduling for Networked Control Systems, *Stefano Longo; Tingli Su; Guido Herrmann; Phil Barber*

Advances in Missile Guidance, Control, and Estimation, *S.N. Balakrishna; Antonios Tsourdos; B.A. White*

End to End Adaptive Congestion Control in TCP/IP Networks,
Christos N. Houmkozlis; George A Rovithakis

Robot Manipulator Control: Theory and Practice, *Frank L. Lewis; Darren M Dawson; Chaouki T. Abdallah*

Quantitative Process Control Theory, *Weidong Zhang*

Classical Feedback Control: With MATLAB® and Simulink®, Second Edition,
Boris Lurie; Paul Enright

Intelligent Diagnosis and Prognosis of Industrial Networked Systems,
Chee Khiang Pang; Frank L. Lewis; Tong Heng Lee; Zhao Yang Dong

Synchronization and Control of Multiagent Systems, *Dong Sun*

Subspace Learning of Neural Networks, *Jian Cheng; Zhang Yi; Jiliu Zhou*

Reliable Control and Filtering of Linear Systems with Adaptive Mechanisms,
Guang-Hong Yang; Dan Ye

Reinforcement Learning and Dynamic Programming Using Function Approximators, *Lucian Busoniu; Robert Babuska; Bart De Schutter; Damien Ernst*

Modeling and Control of Vibration in Mechanical Systems, *Chunling Du; Lihua Xie*

Analysis and Synthesis of Fuzzy Control Systems: A Model-Based Approach,
Gang Feng

Lyapunov-Based Control of Robotic Systems, *Aman Behal; Warren Dixon; Darren M. Dawson; Bin Xian*

System Modeling and Control with Resource-Oriented Petri Nets,
MengChu Zhou; Naiqi Wu

Sliding Mode Control in Electro-Mechanical Systems, Second Edition,
Vadim Utkin; Juergen Guldner; Jingxin Shi

Autonomous Mobile Robots: Sensing, Control, Decision Making and Applications, *Shuzhi Sam Ge; Frank L. Lewis*

Linear Control Theory: Structure, Robustness, and Optimization,
Shankar P. Bhattacharyya; Aniruddha Datta; Lee H.Keel

Optimal Control: Weakly Coupled Systems and Applications, *Zoran Gajic*

Deterministic Learning Theory for Identification, Recognition, and Control,
Cong Wang; David J. Hill

Intelligent Systems: Modeling, Optimization, and Control, *Yung C. Shin; Myo-Taeg Lim; Dobrila Skataric; Wu-Chung Su; Vojislav Kecman*

FORTHCOMING TITLES

Real-Time Rendering: Computer Graphics with Control Engineering,
Gabriyel Wong; Jianliang Wang

Linear Control System Analysis and Design with MATLAB®, Sixth Edition,
Constantine H. Houpis; Stuart N. Sheldon

Modeling and Control for Micro/Nano Devices and Systems,
Ning Xi; Mingjun Zhang; Guangyong Li

Anti-Disturbance Control for Systems with Multiple Disturbances

Lei Guo • Songyin Cao

CRC Press
Taylor & Francis Group
Boca Raton London New York

CRC Press is an imprint of the
Taylor & Francis Group, an **informa** business

CRC Press
Taylor & Francis Group
6000 Broken Sound Parkway NW, Suite 300
Boca Raton, FL 33487-2742

First issued in paperback 2017

© 2014 by Taylor & Francis Group, LLC
CRC Press is an imprint of Taylor & Francis Group, an Informa business

No claim to original U.S. Government works
Version Date: 20130806

ISBN 13: 978-1-138-07668-6 (pbk)
ISBN 13: 978-1-4665-8746-5 (hbk)

Library of Congress Cataloging-in-Publication Data

Guo, Lei, 1961-
     Anti-disturbance control for systems with multiple disturbances / Lei Guo, Songyin Cao.
          pages cm -- (Automation and control engineering ; 52)
     Includes bibliographical references and index.
     ISBN 978-1-4665-8746-5 (hardback)
     1. Nonlinear control theory. I. Cao, Songyin. II. Title.

QA402.35.G86 2013
629.8'36--dc23   2013026469

Visit the Taylor & Francis Web site at
http://www.taylorandfrancis.com

and the CRC Press Web site at
http://www.crcpress.com

Dedication

For our families with love, pride and gratitude

Contents

List of Figures

List of Tables

Foreword

The past several decades have witnessed rapid growth of advanced control applications to many conventional and emerging industries. Anti-disturbance control, as one of the central topics, has been increasingly popular due to its wide applications to aircraft attitude control, high-precision mechatronics and industrial process control. Due to the complexity of the dynamical systems as well as the existence of various sources of disturbances, most of the existing approaches such as H_∞ control and disturbance observer based control (DOBC) cannot be deployed individually to achieve the system performance. Instead, composite hierarchical control structures are attracting more and more attention, thanks to their capability of combining various control schemes and multiple control loops to optimize the system performance against the sophisticated external disturbance environments and the uncertainties of the dynamical model.

In most of the practical control systems, the unknown disturbances can be categorized into Gaussian noise, norm bounded random disturbances, disturbances described by exogenous systems, as well as equivalent disturbances representing the unmodeled dynamics and system uncertainties. As a result, disturbance attenuation and rejection is technically challenging. The situation is further complicated in the case of nonlinear systems with multiple disturbances. Similar challenges exist for the filtering problem, where various types of disturbances and the existence of uncertainties and nonlinearities limit the use of conventional methods such as Kalman filter and the minimum variance filter.

The aim of this monograph is to provide a summary of analysis and synthesis methods on anti-disturbance control and filtering, as well as their application to flexible aircraft attitude control, multi-agent control systems and fault diagnosis. Quite different from most of the existing results, this book puts emphasis on the theory and applications of composite hierarchical anti-disturbance control approaches. In spite of the abundant research results in the general area of anti-disturbance control, systematic approaches on composite hierarchical anti-disturbance control are still relatively rare. It is one of very few books that bridge this gap by systematically reporting important advances in this area, and is certainly a welcome addition and valuable reference for anyone working in this area.

As a leading researcher in this field, the author has successfully delivered some important concepts and systematic approaches in composite hierarchical anti-disturbance control and filtering, with the elegance of mathematics and the relevance of industrial applications. This monograph summarizes the recent advances in anti-disturbance control and filtering and is a must-read book in this area. This book is presented with an easy-to-understand style without loss of technical rigorousness, which makes it very suitable for academic researchers, postgraduate students and industrial engineers who are conducting research in this area, or looking for some guidance in applying advanced anti-disturbance control and filtering methods in their own fields of interests.

University of Manchester, Manchester, UK   *Hong Wang*
March 2013

Preface

Unknown disturbances originating from various sources exist in all practical controlled systems, where unmodeled dynamics and uncertainties can also be formulated as an equivalent disturbance. As such, disturbance attenuation and rejection for nonlinear systems is a challenging objective in the area of control. Analysis and synthesis for nonlinear control systems with disturbances has been one of the most active research fields in the past few decades.

There are several drawbacks to be overcome in future studies on nonlinear anti-disturbance control. First, in engineering applications, the disturbance may originate from various sources and can be described by a composite form rather than a single variable. In this case, the H_∞ control may be too conservative to provide highly accurate control performance. On the other hand, disturbance rejection approaches usually need precise models for both the controlled system and the disturbance system. This confines their applications since the disturbance is also described by a single output variable from a precise exosystem and robustness is difficult to guarantee. For example, although disturbance observer based control is a valid disturbance rejection strategy for nonlinear systems with harmonic disturbances, the performance of the system will deteriorate if the disturbance model cannot be described precisely.

In filtering problems, both external noises, measurement noises and structure vibrations, and also unmodeled, nonlinear and uncertain dynamics are usually merged into the disturbance variable. The accuracy of Kalman filter and minimum variance filtering cannot satisfy this requirement, since the "merged"disturbance is not Gaussian in most applications. On the other hand, robust filtering methods have a large conservativeness for disturbances described by a norm bound variable.

Following funding support from the Chinese National Science Foundation (Grants No. 61203049, 61203195, 61127007, 60935012, 60774013, 60474050), the Chinese 863 project, and the Chinese 973 project (2012CB720003), since 1999 a lot of work has been carried out by the authors using linear matrix inequalities (LMIs) to solve various control design problems encountered in stochastic systems with non-Gaussian noises and multiple disturbances. A number of technical papers have been published in leading control journals and conferences. In fact, different from most previous works, the concept and framework of refined anti-disturbance control are first addressed. A fruitful and systematic results have recently been established. To improve control accuracy and enhance anti-disturbance capability, we need to sufficiently analyze the characteristics of various disturbances based on information measurement. Furthermore, we can formulate models which describe the different types of multiple disturbances. In order to improve accuracy, both disturbance rejection and attenuation performances should be achieved simultaneously for systems with multiple disturbances. In this context, we feel that it would be beneficial to write a book on this particular subject to introduce the recent advances in this control framework for the purpose of enhancing control accuracy.

The book begins with a summary of cutting edge developments in anti-disturbance control, and then focuses on the detailed descriptions of various control and filtering strategies. We hope that readers can obtain a full picture of the latest developments in anti-disturbance control and filtering.

The technical contributors of this book include: Xinjiang Wei from Ludong University, Yantai, China; Xinyu Wen from Taiyuan University of Science and Technology, Taiyuan, China; Hua Liu from Hangzhou Dianzi University, Hangzhou, China; Xiumin Yao from North China Electric Power University, Baoding, China; Hongyong Yang from Ludong University, Yantai, China. The authors would like to thank their families and friends for their consistent support during the writing of the book. In particular, the first author would like to thank his parents (Yitian Guo and Wuqin Qi), his wife (Xiaojie Liu) and his son (Xiaoyu Guo). The second author would like to thank his wife (Qian Zhang) and his son (Zhangyuhan Cao) for their understanding and support.

Beihang University, Beijing, China   *Lei Guo*

Yangzhou University, Yangzhou, China   *Songyin Cao*

Abbreviations and Notations

Throughout this book, the following conventions and notations are adopted:

ACS   Attitude control system

CHADC   Composite hierarchical anti-disturbance control

DOBC   Disturbance observer based control

EKF   Extended Kalman filter

FDD   Fault detection and diagnosis

FTC   Fault-tolerant control

GINS   Gimbaled inertial navigation system

INS   Inertial navigation system

KF   Kalman filter

LMI   Linear matrix inequality

LQG   Linear-quadratic-Gaussian

MIMO   Multiple-input-multiple-output

MJLS   Markovian jump linear system

PDE   Partial differential equation

PDI   Partial differential inequality

PID   Proportional integral derivative

SINS   Strapdown inertial navigation system

SISO   Single-input-single-output

SMC   Sliding mode control

STD   Standard deviation

TSM   Terminal sliding mode

UUB   Uniformly ultimately bounded

UKF   Unscented Kalman filter

VSC   Variable structure control

VSS   Variable structure system

R The field of real numbers

R^n	The set of n-dimensional real vectors or tuples
$I(I_n)$	The identity matrix (of dimension $n \times n$)
$R^{n \times m}$	The set of $n \times m$-dimensional real matrices
max S	The maximum element of set S
min S	The minimum element of set S
sup S	The smallest number of that is larger than or equal to each element of set S
inf S	The largest number of that is smaller than or equal to each element of set S
$[a,b)$	The real number set $\{t \in R : a \le t < b\}$ or the integer set $\{t \in N : a \le t < b\}$
x^T or A^T	The transpose of vector x or matrix A
A^{-1}	The inverse of matrix A
$P > 0$	Matrix P is real symmetric positive definite
$P \ge 0$	Matrix P is real symmetric semi-positive definite
$P < 0$	Matrix P is real symmetric negative definite
$P \le 0$	Matrix P is real symmetric semi-negative definite
$\|x\|$	The Euclidean norm of vector x
$\|A\|$	The induced Euclidean norm of matrix A
$\|x\|_p$	The l_p norm of vector x
$\|A\|_p$	The induced l_p norm of matrix A
$\lambda(A)$	The set of eigenvalues of A
$\lambda_{max}(A)$	The maximum eigenvalues of real symmetric A
$\lambda_{min}(A)$	The minimum eigenvalues of real symmetric A
diag(A)	The diagonal matrix of matrix A
sym(A)	The value of matrix $A + A^T$
tr(A)	The trace of matrix A
$*$	The corresponding elements in the symmetric matrix

1 Developments in CHADC for Systems with Multiple Disturbances

1.1 INTRODUCTION

In a complex environment, there are not only nonlinear dynamic models, uncertain dynamic, time delay, and other unmodeled errors in controlled plants, but also sensor measurement noise, control error and structural vibrations as well as environmental disturbance. The presence of different types of disturbances will seriously affect control accuracy. Therefore, anti-disturbance capability has become a bottleneck problem in control technology. In general, there are several control approaches focusing on nonlinear systems with unknown disturbances:

1. Feedback Linearization Techniques: so that linear robust control techniques can be used [79];
2. Stochastic Nonlinear Control Theory: where disturbances are modelled by a stochastic process, and a nonlinear partial differential equation (PDE) (called Hamilton-Jaccobi-Bellman Equation) has to be solved [2, 7, 91];
3. Nonlinear H_∞ Control: which can attenuate the influences from disturbances to controlled output to a desired level, for systems with bounded disturbances [44, 136, 238];
4. Nonlinear Output Regulation Theory: the solution is based on a Francis-Isidori-Byrnes nonlinear PDE [17, 43, 46, 106, 112, 208, 210];
5. Constructive Nonlinear Control Theory: using a back-stepping procedure [127];
6. Adaptive Nonlinear Control Theory: the adaptive law is adopted for disturbance rejection [3, 143, 144, 146, 159].

Disturbance attenuation, compensation and rejection have been a hot topic in the control field [63, 86, 107, 111, 157, 183, 206, 238], due to the increasing complexity of controlled plants and higher demand for system accuracy, reliability, and real-time performance.

Besides supplying satisfactory stability performance, a desired controller should be simple in structure, fast in computation and robust against variations in parameters, unmodeled dynamics and noise. Some classical control approaches can provide simple design methods but cannot achieve satisfying robust performances. On the contrary, other methods may supply perfect performance, but have overly complicated controller structures or are too dependent on heavy computation to use in practical engineering. Also, some strict assumptions for the models may restrict their applications for different plants.

Feed-forward compensation for modeling errors and exogenous disturbance has been considered as a "robust"control scheme when the errors or disturbance can be measured. However, in most cases they are time-varying unknown dynamics, for which observers or filters are designed. To date, the disturbance observer based control (DOBC) approach has been widely used in practical engineering as a robust method against disturbance [30, 32, 133, 182]. In the first stage, only linear DOBC (LDOBC) is applied to linear models in the frequency domain. Instead of designing a nonlinear control law to compensate for nonlinearities in a dynamical system, a filter or an observer is used to estimate the disturbance where the nonlinearities are treated as a part of disturbances acting on a linear plant. However, even for LDOBC, most results are application-oriented and rigorous theoretical analysis is required. Obviously, a nonlinear DOBC (NLDOBC) may improve the performance and robustness against noises and unmodeled dynamics for some nonlinear systems. To this end, NLDOBC has been applied to robotic systems with constant disturbance and with harmonic disturbances respectively in [4, 29, 58, 31, 124]. The simulations and experiments demonstrated encouraging results.

1.2   DISTURBANCE OBSERVER BASED CONTROL

Consider the following nonlinear dynamic system with unknown disturbances

$$\begin{cases} \dot{x} &= f(x,u,w) \\ y &= h(x,u) \end{cases} \tag{1.1}$$

where $x \in R^n$, $w \in R^{m_1}$, $u \in R^{m_2}$ and $y \in R^{p_1}$ are state, exogenous disturbance, control input and measurement output, respectively. Sometimes w may be described as an exosystem which has a known separate dynamic linear or nonlinear model. The problem to be considered in DOBC theory is to design a nonlinear observer for the nonlinear system (1.1) to estimate the unknown disturbance w, and then design a proper controller to reject the disturbance using its estimation.

1.2.1   LDOBC

The idea of DOBC first appeared in the late 1980s (see for example, [155] and references therein), where an observer was proposed to estimate external disturbance. The estimation of disturbance is immediately supplied to a servo system, so that the disturbance can be compensated for. It is noted that DOBC originated from robotics for independent joint control [155]. Its basic framework was shown originally for linear systems in the frequency domain. Since then, DOBC has been widely used in the mechatronics community as a robust method for disturbance rejection. Effective applications include robot manipulators [10, 24, 57, 123, 133, 155, 162], CNC machine centers in cutting processes [39], high speed direct-drive positioning tables [121], permanent magnet synchronous motors [49], table drive systems [113], hard disks [110], magnetic hard drive servo systems [108], active noise control [13], etc. LDOBC has also been used for friction compensation [58] and for sensorless torque/force control [125]. In LDOBC, all nonlinearities as well as disturbances were included in a generalized disturbance affecting the system in a linear manner, so that linear design techniques could be applied.

1. Frequency Domain Based Approach: Consider the linear system described in Figure 1.1, where u is the control input, d the disturbance, r the reference input, m the measurement noise, y the output and G the controlled plant (see also [162]). In the observer part, G_0 denotes a nominal plant, and $Q(s)$ stands for a low-pass filter to be determined, which is the main task in the disturbance observer design.

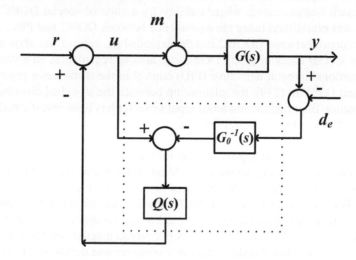

FIGURE 1.1   Disturbance observer for linear systems in the frequency domain.

For causality, the relative degree of Q should be no less than that of the transfer function describing the nominal plant. The closed-loop system can be obtained as

$$y(s) = \{Q(s)G_0^{-1}(s) + [I - Q(s)]G^{-1}(s)\}^{-1}$$
$$\{r - Q(s)G_0^{-1}(s)m - [I - Q(s)]de\} \tag{1.2}$$

where the equivalent disturbance is given by

$$d_e = [G^{-1}(s) - G_0^{-1}(s)]y + d - G_0^{-1}(s)m \tag{1.3}$$

It is shown that the disturbance observer can estimate both d and the nonlinearity, the parameter variation and the measurement noise. Therefore, within a specified frequency range $Q(s) \approx I$ the real plant of $G(s)$ will behave as $G_0(s)$ in spite of the presence of d. Hence, this is a very simple and direct approach to reject the disturbance.

Along with this framework, there are many works devoted to various practical plants especially for robot manipulators which are well known as linearizable nonlinear systems. Besides the references listed above, one can also see [125, 133, 161, 162].

However, generally $Q(s)$ is sensitive to the sensor noise, and a proper cut-off frequency together with $Q(s)$ should be selected so that the sensor noise does not grow excessively. Another shortcoming is that less systematic stability analysis results can be provided if a pure frequency domain approach is used.

In [133], the root-locus method was used to verify the stability of DOBC. In [10], a comparison between DOBC and the passivity-based control (PBC) approach was presented, where stability for a class of special DOBC design was established using the equivalence between DOBC and PBC. The equivalence between LDOBC and the extended H_∞ control was shown for some special cases in [150]. DOBC can also be regarded as an extended proportional integral derivative (PID) control in the disturbance rejection context [179]. In [179], the relationship between the so-called disturbance estimation filter and the unknown input observer has been investigated for linear plants in the frequency domain.

2. Time Domain Based Approach: To overcome the shortcomings of the frequency domain based approach, time domain based approaches have been provided in a series of publications. Most of them also focused on linear models or some linearizable nonlinear systems such as robot manipulators. For example, one can see [150] for the time-invariant unknown disturbance and [162] for the time-varying unknown disturbance. The approach in [49] is a typical LDOBC scheme for permanent magnet synchronous motors, where a reduced-order disturbance observer was developed. In [133], LDOBC was used to decouple multiple-input-multiple-output (MIMO) systems, where the stability analysis of LDOBC was provided based on the root-locus method.

In [35], a disturbance observer method based on variable structure systems (VSS) was proposed for minimum-phase dynamic systems with arbitrary relative degree. Firstly the model uncertainties and nonlinearities as well as some exogenous inputs are merged into the disturbance term and are regarded as a part of the disturbances. Then a variable structure control (VSC) can be designed for the new linear model with new disturbances to estimate them.

As stated above, besides the linear plants, some nonlinear models also can be dealt with by LDOBC, where noise, unmodeled dynamics and nonlinearities can be merged into a new equivalent disturbance. In this case, when the new disturbance is regarded as a part of a state, the (reduced-order) observer design theory can be applied and stability analysis can be easily established.

1.2.2 NLDOBC

Although LDOBC algorithms have been successfully applied in industrial engineering, it is expected that if the nonlinear dynamics of a model is utilized sufficiently, the performance and the robustness will be improved greatly with the NLDOBC law. Furthermore, for many practical plants where nonlinearities are highly coupled, the LDOBC method is unavailable [36, 138, 221, 218]. Here NLDOBC means that the observer is designed for a disturbance acting on nonlinear systems, and the compensation for the disturbance also acts on nonlinear models. Up to now, a few NLDOBC schemes have been proposed for some nonlinear systems, which will be introduced in the following paragraphs.

1. Nonlinear Observer Design Approach: as in the linear case, nonlinear observer theory can be used if disturbances are considered as a part of states. In [65], a nonlinear observer for Coulomb friction has been presented. Generally speaking, this approach mainly depends on (and developed along with) the nonlinear observer theory, as shown in [49] for linear systems. In many cases, an observer structure is not very easy to realize in engineering especially for complex plants.

   To shorten the gap between linear and nonlinear disturbance observer, a robust result is desired for systems with uncertainties and bounded nonlinear dynamics. In [85], several robust DOBC schemes have been provided, where the models with both known and unknown nonlinear dynamics have been discussed. The separation principle can be verified for some cases of robust DOBC schemes. Simulations for a helicopter show that better robustness can be achieved against parameter uncertainties, compared with a classical robust control law [167].

2. NLDOBC Based on Lyapunov Theory: when the disturbance is supposed to satisfy special conditions such as being a constant, it can possibly be estimated via measurement without an observer. In [58], an NLDOBC law was presented for a class of multivariate nonlinear systems including two link robot manipulators. Global stability was guaranteed and the encouraging experimental results show that the performance and robustness against friction and uncertain dynamics is greatly improved. Furthermore, for a class of systems with unified relative degree under time-invariant disturbances, a new nonlinear DOB control scheme was proposed in [57]. Global stability was also established using Lyapunov theory. It is shown that the explicit observer structure disappears and a nonlinear PID controller appears. In [137], for time-invariant disturbances, the estimation of disturbance can be obtained without the normal observer type by using Lyapunov theory. For the tracking control problem of robot manipulators, simulation results show that the DOBC can achieve superior performance against adaptive control schemes and sliding mode control.

   Up to now, most rigorous analysis and synthesis of NLDOBC only concerned time-invariant disturbance. In [29], the regulation problem for a nonlinear system with harmonic disturbances was studied using NLDOBC. The proposed nonlinear DOB controller includes a nonlinear disturbance observer and a conventional controller stabilizing the plant without disturbances. Instead of the Lyapunov theory, passivity theory was applied to establish the L_2 stability for the whole composite controller. Recently, a new NLDOBC scheme has been provided to a class of nonlinear systems with time varying and model-free disturbance [27, 33, 219]. [98] discussed the unknown input observer (or extended state observer) and its application in DOBC. The performance of the observer can be guaranteed by analyzing the estimation error of the dynamic equation. The changing rate of the disturbance is supposed to be bounded and input-to-state stability can be guaranteed. Due to the lack of analysis tools, introducing nonlinearity into control algorithms brings difficulties to convergence and stability analysis [68].

Compared with other existing control methodologies for systems with unknown disturbances, DOBC has several features:

- It is possible to integrate other control strategies with DOBC. This approach provides a framework to enhance disturbance rejection by combining other (nonlinear) control methods that focus on systems in the absence of disturbances. Due to the rich structures of nonlinear systems and functions, NLDOBC can provide much more freedom for controller design.
- DOBC can provide a "real"stability analysis and synthesis approach for plants in the presence of disturbances. Unknown disturbances are modelled more flexibly in DOBC. The design procedure is relatively simple and easy to accept by engineers. And, generally no PDEs or partial differential inequalities (PDIs) have to be solved for controller design.

Although DOBC has been widely used in engineering and has been developed by many researchers, less feasible design methodologies have been provided for NLDOBC of nonlinear systems with unknown disturbances. One challenging research objective is to design NLDOB controllers for more general nonlinear systems, for example, nonminimum-phase systems, nonlinear systems without well-defined relative degree, and MIMO nonlinear systems. Another potential direction is to study the more general disturbance (driven by more general exosystems) or even "model-free" disturbance.

1.3 COMPOSITE HIERARCHICAL ANTI-DISTURBANCE CONTROL

Most of the above-mentioned disturbance attenuation and rejection methods were used for systems with a single disturbance. In practical engineering, complex systems were often described as mathematical models with multiple disturbances [94, 204]. Fox example, disturbances of spacecraft can be divided into three types, including equivalent disturbance caused by modeling error, internal disturbance force torque and external space environmental disturbance force torque. For the navigation and control models of satellites, variations in rotational inertia and coupling character will lead to structural perturbation, uncertainty and unmodeled dynamics. These factors constitute a system modeling error, which is often regarded as an equivalent disturbance. The internal disturbance torque of the satellite can be divided into sensor measurement noise, control actuator error (including flywheel component disturbance, propulsion system error, etc.) as well as structural influence (including solar array flexible attachment-induced structural perturbations and vibration, etc.). Space disturbance torque includes gravity gradient torque, sunlight pressure torque, aerodynamic torque and geomagnetic torque. The magnitude of these torques mainly depends on the orbital altitude of the spacecraft, surface characteristics, atmospheric density, onboard magnets, and so on. In conclusion, spacecraft systems in a complex environment contain not only nonlinear dynamics, uncertainties, coupling, delay, and unmodeled errors, but also internal sensor measurement noise, control error and structural vibration, as well as external space environment disturbance torque [185, 194]. From the viewpoint of control theory, these disturbances and noises can be characterized as an uncertain norm-bounded variable, a harmonic, step signal, non-Gaussian/ Gaussian random variable, a bounded change rate variable, output variables of a neutral stable system, and other types of disturbances.

Currently, anti-disturbance control theory has the following problems. Firstly, the modeling theory was limited to the controlled plant without considering the model and characteristic information of external and internal disturbances. Thus, it reduced control precision. With the development of sensor technology, much attention has been paid to the characteristic analysis of noise and disturbance [53]. However, most of the anti-disturbance control schemes have not combined the characteristic analysis of noise and disturbance, which hampered the application of anti-disturbance control theory. Secondly, the classical anti-disturbance control theory only considered one "equivalent"disturbance which was a combination of different unknown sources. However, it is noted that along with the development of sensor technology and data processing, it has been shown that multiple disturbances exist in most practical processes and can be formulated into different mathematical descriptions after modeling and error analysis. That is, in practical engineering, disturbance originating from different sources can be described separately and formulated into a composite form, rather than a single variable. For example, the H_∞ control may be too conservative to provide highly accurate control performance. On the other hand, disturbance rejection approaches usually need precise models for both the controlled system and the disturbance.

For systems with multiple disturbances, the theoretical bottleneck problem is that different disturbance attenuation and coupling compensators increase the complexity of closed-loop control systems. The stability of the system and disturbance attenuation performance based on a single controller no longer holds. DOBC strategies appeared in the late of 1980s and have been applied in many control areas (see [85, 87, 201], and references therein). For nonlinear systems with multiple disturbances, the H_∞ and variable structure control has been integrated with DOBC in [22, 201, 202]. In [22, 202], multiple disturbances were divided into a norm bounded variable and uncertain modeled disturbance, where DOBC was applied to reject the modeled disturbance and robust H_∞ was used to attenuate the norm bounded disturbance. In [93], a composite DOBC and adaptive control approach was proposed for a class of nonlinear system with uncertain modeled disturbance and the disturbance signal represented by an unknown parameter function. Because of considering the different disturbance characteristics, the above-mentioned composite hierarchical control method can be considered as a composite hierarchical anti-disturbance control (CHADC) method.

The core of the anti-disturbance control is to research the composite hierarchical anti-disturbance control schemes. Its main feature is to make full use of disturbance characteristics, and model the disturbance with different classifications. Then, it can achieve the disturbance attenuation and rejection performance simultaneously. The inner layer of the CHADC includes a disturbance observer and compensator. Meanwhile, the outer layer of the CHADC includes disturbance attenuation controller, such that the hierarchical structure can simplify the design methods, and improve the accuracy of the controller. At present, a theoretical system of DOBC combining with a PID, adaptive, variable structure has been formed. In addition, the idea of CHADC has been applied to the problem of filtering, fault detection and diagnosis for systems with multiple disturbances [19, 20, 21, 22, 82, 170]. It can be predicted that the study of CHADC for systems with multiple disturbances has important theoretical significance and application. The following problems need to be studied in depth on the basis of existing results:

- To analyze the mechanism of disturbance characteristics and disturbance modeling for different objects.
- To relax the restrictions on the disturbance model matching conditions for constructing the disturbance observer.
- To reduce the conservativeness of disturbance and uncertain parameter estimation.
- To introduce the disturbance attenuation and rejection capability for unknown parameters and uncertain disturbances in the existing CHADC schemes.

2 Disturbance Attenuation and Rejection for Nonlinear Systems via the DOBC Approach

2.1 INTRODUCTION

Analysis and synthesis of nonlinear dynamic systems with disturbances has been one of the most active research areas in the past decades. It has been shown that classical control approaches can provide relatively simple design algorithms to deal with disturbances, but lack sound theoretical justifications. On the other hand, other approaches are rigorous mathematically but only suitable for systems with specific structures, or demand a heavy computation burden in applications. For example, partial differential equations (PDEs) involved in nonlinear output regulation theory, stochastic nonlinear control theory and nonlinear H_∞ control are generally difficult to solve. In addition, some approaches are only concerned with the stability of the nominal system in the absence of disturbances, implying that in this case stability of the system cannot be guaranteed in the presence of disturbances (see, e.g., [144, 186]).

In most cases only linear disturbance observer based control (LDOBC) approaches are applied even where a plant possesses strong nonlinearity, where nonlinear dynamics is treated as part of disturbance attached to a linear plant [83]. However, the combination with nonlinear dynamics will change the original properties for some kinds of disturbances such as constant and harmonic. Obviously, for a nonlinear system, a nonlinear DOBC (NLDOBC) law can improve performance and robustness greatly against noise and unmodeled dynamics. To provide satisfactory control performances, new NLDOBC schemes have been proposed and applied to robotic systems [28, 31]. In these results only a class of single-input-single-output (SISO) nonlinear systems with well-defined disturbance relative degree are studied, and constant or harmonic disturbances are involved. In addition, in most existing DOBC approaches, full and precise information for a plant is required in disturbance observer design while in many practical cases only part of the information is available, such as the bound of nonlinearity or part of the states. Generally speaking, up to now systematic NLDOBC strategies are required to develop for more general control plants and wider external disturbances.

This chapter considers the DOBC approaches for a class of multiple-input-multiple-output (MIMO) nonlinear systems. The nonlinear dynamics is described by known and unknown nonlinear functions, respectively, and the disturbances which are not confined to being constant, harmonic or neutral stable are represented by a linear exogenous system. After reformulating the DOBC design problem for the known nonlinearity case, two linear matrix inequality (LMI)-based design schemes are proposed on full-order and reduced-order disturbance observers, respectively. For the uncertain nonlinearity case, a robust DOBC approach is developed to enhance robust performance. By using disturbance estimation, DOBC strategies can be integrated with conventional stabilization controllers to reject disturbance and globally stabilize closed-loop systems. Systematic stability analysis is established using Lyapunov theory and design procedures are proposed based on convex optimization algorithms. Finally, simulations on an A4D aircraft model show the effectiveness of the proposed approaches.

2.2 PROBLEM STATEMENT

Consider a MIMO dynamical system with nonlinearity and disturbances in both state and output equations as follows,

$$
\begin{cases}
\dot{x}(t) & = A_0 x(t) + F_{01} f_{01}(x(t), t) + B_0[u(t) + d(t)] \\
y(t) & = C_0 x(t) + F_{02} f_{02}(x(t), t) + D_0 d(t)
\end{cases}
\tag{2.1}
$$

where $x(t) \in R^n, d(t) \in R^m, u(t) \in R^m$ and $y(t) \in R^{p_1}$ are state, unknown disturbance, control input and measurement output, respectively. $A_0, B_0, C_0, D_0, F_{01}$ and F_{02} are given system matrices. $f_{01}(x(t), t)$ and $f_{02}(x(t), t)$ are nonlinear functions which are supposed to be either known (see Sections 2.3 and 2.4) or unknown (see Section 2.5), but satisfy bounded conditions described as Assumption 2.1.

Assumption 2.1 *For any $x_j \in R^n$, $j = 1, 2$, nonlinear functions $f_{0i}(x(t), t)$ $(i = 1, 2)$ satisfy*

$$
\begin{cases}
\|f_{0i}(x_1(t), t) - f_{0i}(x_2(t), t)\| \le \|U_i(x_1(t) - x_2(t))\| \\
f_{0i}(0, t) = 0, i = 1, 2
\end{cases}
\tag{2.2}
$$

where U_i $(i = 1, 2)$ are given constant weighting matrices.

Remark 2.1 A variety of nonlinear systems can be described by model (2.1). A system with weak nonlinearity certainly can be represented by this model. For a system with strong nonlinearity, when some nonlinear control techniques such as feedback linearization, dynamic inversion control, gain scheduling technique, or some equivalent transformations are applied, the majority of the nonlinearity can be cancelled and the resulting system can also be described by this model. Actually, the research on this model was also motivated by our work on applying DOBC in robotics where the computed torque control (CTC) was designed to linearize a manipulator (see, e.g., [31]). However, due to the variation of the tip mass (load) of the manipulator, only the majority of the nonlinearity in dynamics can be cancelled by CTC. The remaining nonlinearity satisfies some bounded conditions as in Assumption 2.1. Therefore the manipulator under this control law can be well described by system (2.1) with Assumption 2.1. A similar situation also occurs in dynamic inversion control of aerospace systems [30].

Similar to the output regulation theory [111], the unknown external disturbance $d(t)$ is supposed to be generated by an exogenous system described by

$$\begin{cases} \dot{w}(t) &= Ww(t) \\ d_0(t) &= Vw(t) \end{cases} \tag{2.3}$$

Many kinds of disturbances in engineering can be described by this model, for example, unknown constant and harmonics with unknown phase and magnitude [28, 111]. In most existing results the disturbances are restricted to bounded exogenous signals [111, 234], while the following assumption is required so that the control problem is well posed.

Assumption 2.2 (A_0, B_0) *is controllable and* (W, B_0V) *is observable.*

Nonlinear DOBC problem: design the observer for nonlinear system (2.1) to estimate the unknown disturbance $d(t)$, and then construct a composite controller with the observer and a conventional controller so that the disturbance can be rejected and the stability of the resulting composite system can be guaranteed.

2.3 FULL-ORDER OBSERVERS FOR KNOWN NONLINEARITY

In Sections 2.3 and 2.4, we suppose that $f_{0i}(x(t),t)$ $(i=1,2)$ are given and Assumptions 2.1 and 2.2 hold. In Section 2.3, the full state is supposed to be unavailable and needs to be estimated. It will be shown that the considered DOBC problem can be cast into an augmented state estimation problem for systems with nonlinearity. For this purpose, Assumption 2.3 is needed for full-order cases discussed in this section.

Assumption 2.3 (A,C) *is controllable.*

By augmenting state Equation (2.1) with disturbance dynamics Equation (2.3), the composite system is given by

$$\begin{cases} \dot{z}(t) &= Az(t) + F_1 f_1(z(t),t) + Bu(t) \\ y(t) &= Cz(t) + F_2 f_2(z(t),t) \end{cases} \tag{2.4}$$

where

$$z(t) = \begin{bmatrix} x(t) \\ w(t) \end{bmatrix}, A = \begin{bmatrix} A_0 & B_0V \\ 0 & W \end{bmatrix}$$

$$B = \begin{bmatrix} B_0 \\ 0 \end{bmatrix}, F_1 = \begin{bmatrix} F_{01} \\ 0 \end{bmatrix}$$

$$C = \begin{bmatrix} C_0 & D_0 V \end{bmatrix}, f_1(z(t),t) = f_{01}(x(t),t),$$

$$F_2 = F_{02}, f_2(z(t),t) = f_{02}(x(t),t)$$

The full-order observer for both $x(t)$ and $w(t)$ is designed as

$$\begin{cases} \dot{\hat{z}}(t) &= A\hat{z}(t) + F_1 f_1(\hat{z}(t),t) + Bu(t) + L(\hat{y}(t) - y(t)) \\ \hat{y}(t) &= C\hat{z}(t) + F_2 f_2(\hat{z}(t),t) \end{cases} \quad (2.5)$$

where

$$\hat{z}(t) = \begin{bmatrix} \hat{x}(t) \\ \hat{w}(t) \end{bmatrix}, L = \begin{bmatrix} L_1 \\ L_2 \end{bmatrix},$$

$$f_1(\hat{z}(t),t) = f_{01}(\hat{x}(t),t), f_2(\hat{z}(t),t) = f_{02}(\hat{x}(t),t)$$

and L is the observer gain to be determined. The estimation error

$$e(t) = z(t) - \hat{z}(t) = \begin{bmatrix} e_x(t) \\ e_w(t) \end{bmatrix} = \begin{bmatrix} x(t) - \hat{x}(t) \\ w(t) - \hat{w}(t) \end{bmatrix} \quad (2.6)$$

is governed by

$$\begin{aligned} \dot{e}(t) &= (A - LC)e(t) + F_1[f_1(z(t),t) - f_1(\hat{z}(t),t)] \\ &\quad + LF_2[f_2(z(t),t) - f_2(\hat{z}(t),t)] \end{aligned} \quad (2.7)$$

In the DOBC scheme, the control can be constructed as

$$u(t) = -\hat{d}(t) + K\hat{x}(t) \quad (2.8)$$

where the disturbance estimation to compensate $d(t)$ is given by

$$\hat{d}(t) = V\hat{w}(t) = \begin{bmatrix} 0 & V \end{bmatrix} \hat{z}(t) \quad (2.9)$$

and K is the conventional control gain for stabilization.

Combining estimation error Equation (2.7) with plant (2.1) yields

$$\begin{bmatrix} \dot{x}(t) \\ \dot{e}_w(t) \end{bmatrix} = \begin{bmatrix} A_0 + B_0 K & \bar{B} \\ 0 & A - LC \end{bmatrix} \begin{bmatrix} x(t) \\ e_w(t) \end{bmatrix}$$

$$+ \begin{bmatrix} F_{01} & 0 & 0 \\ 0 & F_1 & LF_2 \end{bmatrix} \begin{bmatrix} f_{01}(z(t),t) \\ f_{01}(x(t),t) - f_{01}(\hat{x}(t),t) \\ f_{02}(x(t),t) f_{02}(\hat{x}(t),t) \end{bmatrix} \quad (2.10)$$

where $B = \begin{bmatrix} -B_0 K & B_0 V \end{bmatrix}$. It can be verified that

$$\left\| \begin{bmatrix} f_{01}(z(t),t) \\ f_{01}(x(t),t) - f_{01}(\hat{x}(t),t) \\ f_{02}(x(t),t) f_{02}(\hat{x}(t),t) \end{bmatrix} \right\| \leq \left\| \begin{bmatrix} U_1 & 0 & 0 \\ 0 & U_2 & 0 \\ 0 & U_3 & 0 \end{bmatrix} \begin{bmatrix} x(t) \\ e_x(t) \\ e_w(t) \end{bmatrix} \right\| \quad (2.11)$$

At this stage, our objective is to find L and K such that system (2.10) is asymptotically stable. For the sake of simplifying descriptions, we denote

$$\bar{F}_1 = [\ \lambda_1 P_2 F_1 \quad \lambda_2 R_2 F_2\ ], \bar{U}_1 = [\ U_1 \quad 0\ ], \bar{U}_2 = [\ U_2 \quad 0\ ] \qquad (2.12)$$

Theorem 2.1

Consider system (2.1) with disturbances (2.3) under Assumptions 2.1-2.3. For some $\lambda_1 > 0$ and $\lambda_2 > 0$, if there exist $Q_1 > 0$ and R_1 satisfying

$$\begin{bmatrix} sym(A_0 Q_1 + B_0 R_1) + \lambda_1^2 F_{01} F_{01}^T & \frac{1}{\lambda_1} Q_1 U_1^T \\ * & -I \end{bmatrix} < 0 \qquad (2.13)$$

and $P_2 > 0$ and R_2 satisfying

$$\begin{bmatrix} sym(P_2 A - R_2 C) + \frac{1}{\lambda_1} \bar{U}_1^T \bar{U}_1 + \frac{1}{\lambda_2} \bar{U}_2^T \bar{U}_2 & \bar{F}_1 \\ * & -I \end{bmatrix} < 0 \qquad (2.14)$$

then closed-loop system (2.10) under DOBC law (2.8) with gain $K = R_1 Q_1^{-1}$ and observer (2.5) with gain $L = P_2^{-1} R_2$ is asymptotically stable. ∎

Proof. Let

$$\Phi_1(x,t) = x^T(t) Q_1^{-1} x(t) + \frac{1}{\lambda_1^2} \int_0^t [\|U_1 x(\tau)\|^2 - \|f_{01}(x,\tau)\|^2] d\tau$$

and

$$\begin{aligned} \Phi_2(e,t) &= e^T(t) P_2 e(t) + \frac{1}{\lambda_1^2} \int_0^t [\|U_1 e_x(\tau)\|^2 - \|f_1(z,\tau) - f_1(\hat{z},\tau)\|^2] \\ &\quad + \frac{1}{\lambda_2^2} \int_0^t [\|U_2 e_x(\tau)\|^2 - \|f_2(z,\tau) - f_2(\hat{z},\tau)\|^2] d\tau \end{aligned}$$

Following (2.12) and (2.13), it is noted that for all x, e, $\Phi_1(x,t)$, Along with the trajectories of (2.10), firstly it can be verified that

$$\begin{aligned} \dot{\Phi}_2(e,t) &= e^T(t) sym[P_2(A-LC)]e(t) + 2e^T P_2 [F_1(f_1(z,t) - f_1(\hat{z},t)) \\ &\quad + LF_2(f_2(z,t) - f_2(\hat{z},t))] + \frac{1}{\lambda_1^2}[\|U_1 e_x(t)\|^2 - \|f_1(z,t) - f_1(\hat{z},t)\|^2] \\ &\quad + \frac{1}{\lambda_2^2}[\|U_2 e_x(t)\|^2 - \|f_2(z,t) - f_2(\hat{z},t)\|^2] \\ &\leq e^T(t) sym[P_2(A-LC)]e(t) + e^T(t)[\lambda_1^2 P_2 F_1 F_1^T P_2 \\ &\quad + \lambda_2^2 P_2 LF_2 F_2^T L^T P_2 + \frac{1}{\lambda_1^2} \bar{U}_1^T \bar{U}_1 + \frac{1}{\lambda_2^2} \bar{U}_2^T \bar{U}_2] e(t) \\ &\leq -\eta_1 \|e(t)\|^2 \end{aligned}$$

which can be guaranteed by (2.13) using Schur complement, where $\eta_1 > 0$ is a proper constant. Similarly, for (2.10) in the absence of e, based on (2.12) we can find a proper constant η_2 such that

$$
\begin{aligned}
\Phi_1(x,t) &= x^T(t)[P_1(A_0 + B_0 K) + (A_0 + B_0 K)^T P_1 \\
&\quad + \lambda_1^2 P_1 F_{01} F_{01}^T P_1 + \frac{1}{\lambda_1^2} U_1^T U_1] x(t) \\
&\leq -\eta_2 \|e(t)\|^2
\end{aligned}
$$

where $P_1 = Q_1^{-1}$, which also means the pure stabilization problem for (2.1) without disturbances d is solvable. If both (2.12) and (2.13) hold, then there exists $\eta_3 > 0$, depending on P_1 and $\bar{B}$, such that $2x^T(t)P_1 \bar{B} e(t) \leq \|x(t)\|\|e(t)\|$.

A Lyapunov function candidate for (2.10) is chosen as

$$
\Phi(x,e,t) = \Phi_1(x,t) + \eta_0 \Phi_2(e,t)
$$

where $\eta_0 = \eta_3^2 / 4\eta_1 \eta_2$. Thus, along with closed-loop system (2.10), we have

$$
\begin{aligned}
\Phi(x,e,t) &\leq -\eta_0 \eta_1 \|e\|^2 - \eta_2 \|x\|^2 + 2x^T P_1 \bar{B} e \\
&\leq -\eta_0 \eta_1 \|e\|^2 - \eta_2 \|x\|^2 + \eta_3 \|x\|\|e\| \\
&= -(\sqrt{\eta_2}\|x\| + \frac{\eta_3}{2\sqrt{\eta_2}}\|e\|)^2 \\
&\leq -\min\{\sqrt{\eta_2}, \frac{\eta_3}{2\sqrt{\eta_2}}\} \left\| \begin{bmatrix} x \\ e \end{bmatrix} \right\|^2
\end{aligned}
$$

Correspondingly the closed-loop system is asymptotically stable.   □

Remark 2.2 The observer and controller design can be obtained separately as follows (see, e.g., [81]):

1. compute the observer gain L via (2.13) together with P_2 and R_2;
2. solve (2.12) to obtain Q_1, R_1 and K;
3. construct the control law based on (2.8).

Remark 2.3 The choices of parameters λ_i ($i = 1, 2$) in Theorem 2.1 (also λ in Theorems 2.2 and 2.3) are involved in amplification of the inequality related to the differential of Lyapunov function. Such parameters will have influence on determination of the controller gain and the convergent rates of the states, which generally should be selected by trial and error based on concrete situations. Generally, there is a trade-off among high gain, solvability and convergence.

2.4 REDUCED-ORDER OBSERVERS FOR KNOWN NONLINEARITY

When all states of a system are available, it is unnecessary to estimate the states, and only the disturbance needs to be estimated. In this section, the disturbance observer is formulated as

$$\begin{cases} \hat{d}(t) & = & V\hat{w}(t) \\ \hat{w}(t) & = & v(t) - Lx(t) \end{cases} \tag{2.15}$$

where $v(t)$ is the auxiliary variable generated by

$$\begin{aligned} \dot{v}(t) & = & (W + LB_0V)(v(t) - Lx(t)) - L[A_0x(t) \\ & & -B_0u(t) - F_{01}f_{01}(x(t),t)] \end{aligned} \tag{2.16}$$

Comparing (2.1) and (2.3) with (2.15) yields

$$\dot{e}_w(t) = (W + LB_0V)e_w(t) \tag{2.17}$$

When DOBC law $u(t) = -\hat{d}(t) + Kx(t)$ is applied, the composite system combined with (2.15) is given by

$$\begin{bmatrix} \dot{x}(t) \\ \dot{e}_w(t) \end{bmatrix} = \begin{bmatrix} A_0 + B_0K & B_0V \\ 0 & W + LB_0V \end{bmatrix} \begin{bmatrix} x(t) \\ e_w(t) \end{bmatrix}$$

$$+ \begin{bmatrix} F_{01} \\ 0 \end{bmatrix} f_{01}(x(t),t) \tag{2.18}$$

In this case, it is obvious that the observer error dynamics is separated from the controller design.

Theorem 2.2

Consider system (2.1) with the disturbances (2.3) under Assumptions 2.1 and 2.2. For some $\lambda > 0$, if there exist $Q_1 > 0$ and R_1 satisfying

$$\begin{bmatrix} sym(A_0Q_1 + B_0R_1) + \lambda^2 F_{01}F_{01}^T & \frac{1}{\lambda}Q_1U_1^T \\ * & -I \end{bmatrix} < 0 \tag{2.19}$$

and $P_2 > 0$ and R_2 satisfying

$$sym(P_2W + R_2B_0V) < 0 \tag{2.20}$$

then closed-loop system (2.18) under DOBC law (2.8) with gain $K = R_1Q_1^{-1}$ and observer (2.15) with gain $L = P_2^{-1}R_2$ is asymptotically stable.   ∎

   Proof. Firstly, we denote

$$P = \begin{bmatrix} P_1 & 0 \\ 0 & P_2 \end{bmatrix} = \begin{bmatrix} Q_1^{-1} & 0 \\ 0 & P_2 \end{bmatrix} > 0 \tag{2.21}$$

A Lyapunov function candidate is chosen as

$$\Sigma = \begin{bmatrix} x^T(t) & e_w^T(t) \end{bmatrix} P \begin{bmatrix} x^T(t) & e_w^T(t) \end{bmatrix}^T$$

Similar to the proof of Theorem 2.1, it can be verified that along with the trajectories of (2.18), $\dot{\Sigma} < 0$ holds if there exists $P > 0$ satisfying

$$\begin{bmatrix} sym(\Theta_0) & \lambda P F_1 & \lambda^{-1} U_1^T \\ * & -I & 0 \\ * & * & -I \end{bmatrix} < 0 \qquad (2.22)$$

where

$$\Theta_0 = P \begin{bmatrix} A_0 + B_0 K & B_0 \\ 0 & W + LB_0 \end{bmatrix}, F_1 = \begin{bmatrix} F_{01} \\ 0 \end{bmatrix}$$

If P is chosen as (2.21), then (2.22) holds if

$$\Pi_1 - P_1 B_0 \Pi_2^{-1} B_0^T P_1 < 0 \qquad (2.23)$$

based on Schur complement, where

$$\Pi_1 = sym(P_1(A_0 + B_0 K)) + \lambda^2 P_1 F_{01} F_{01}^T P_1^T + \frac{1}{\lambda^2} U_1^T U_1,$$

$$\Pi_2 = sym(P_2(W + LB_0 V))$$

On the other hand, it can be shown that (2.19) and (2.20) are equivalent to $\Pi_1 < 0$ and $\Pi_2 < 0$, respectively. $\Pi_1 < 0$ means there exist $Q_1 > 0$ and $\alpha > 0$ such that

$$sym[(A_0 + B_0 K)Q_1] + \lambda^2 F_{01} F_{01}^T + \lambda^{-2} Q_1 U_1^T U_1 Q_1 < -\alpha I \qquad (2.24)$$

When $\Pi_2 < 0$, two constants $\beta_i > 0$ $(i = 1, 2)$ can be obtained such that $-\beta_1 I < \Pi_2 < -\beta_2 I$ for a given P_2. Thus, it can be selected $\tilde{P}_2 = \delta P_2 (\delta > 0)$ where δ is sufficiently large such that $\delta^{-1} \beta_1^{-1} B_0 B_0^T < \alpha I$. Noting (2.23) holds if

$$sym[(A_0 + B_0 K)Q_1] + \lambda^2 F_{01} F_{01}^T + \lambda^{-2} Q_1 U_1^T U_1 Q_1 + \beta_1^{-1} B_0 B_0^T < 0$$

the proof is completed after P_2 is replaced by $\tilde{P}_2$. $\square$

Remark 2.4 One of the features of the proposed DOBC for linear systems is that the disturbance observer can be designed separately from the controller design. Theorems 2.1 and 2.2 extend the separation principle for linear systems to a class of nonlinear systems.

2.5 REDUCED-ORDER OBSERVERS FOR UNKNOWN NONLINEAR-ITY

In this section, it is supposed that Assumptions 2.1 and 2.2 also hold but the nonlinear functions in (2.15) are unknown. Different from Sections 2.3 and 2.4, in this case $f_{01}(x(t),t)$ is unavailable in observer design. To save space, only reduced-order observer design is considered.

In this case, the disturbance observer also can be formulated as (2.15), while the auxiliary dynamics is governed by

$$\dot{v}(t) \quad = \quad (W + LB_0V)(v(t) - Lx(t)) - L(-A_0x(t) - B_0u(t)) \qquad (2.25)$$

rather than (2.16), Estimation error $e_w(t) = w(t) - \hat{w}(t)$ satisfies

$$\dot{e}_w(t) = (W + LB_0V)e_w(t) - LF_{01}f_{01}(x(t),t) \qquad (2.26)$$

Hence, the composite system with (2.25) and (2.8) is given by

$$\begin{bmatrix} \dot{x}(t) \\ \dot{e}_w(t) \end{bmatrix} = \begin{bmatrix} A_0 + B_0K & B_0V \\ 0 & W + LB_0V \end{bmatrix} \begin{bmatrix} x(t) \\ e_w(t) \end{bmatrix}$$

$$+ \begin{bmatrix} F_{01} \\ -LF_{01} \end{bmatrix} f_{01}(x(t),t) \qquad (2.27)$$

Different from Section 2.4, the error dynamic equation has uncertainty and the system matrix of the closed-loop system described by (2.27) does not have a triangular structure.

Theorem 2.3

Consider system (2.1) with the disturbances (2.3) under Assumptions 2.1 and 2.2. For some $\lambda > 0$, if there exist $\alpha_1 > 0$, $Q_1 > 0$ and R_1 satisfying

$$\begin{bmatrix} sym(A_0Q_1 + B_0R_1) + \lambda^2 F_{01}F_{01}^T + \alpha_1 I & \frac{1}{\lambda}Q_1U_1^T \\ * & -I \end{bmatrix} < 0 \qquad (2.28)$$

and $\alpha_2 > 0$, $P_2 > 0$ and R_2 satisfying

$$sym(P_2W + R_2B_0V) + \alpha_2 I < 0 \qquad (2.29)$$

together with constraint

$$\gamma_1\gamma_2 < \alpha_1\alpha_2 \qquad (2.30)$$

where $\gamma_1 = 2\|Q_1^{-1}B_0V\|$, $\gamma_2 = 2\|R_2F_{01}\|\|U_1\|$, then closed-loop system (2.27) under DOBC law (2.8) with gain $K = R_1Q_1^{-1}$ and observer (2.15) with gain $L = P_2^{-1}R_2$ is asymptotically stable. ∎

Proof. Denote

$$\Gamma_1(x,t) = x^T(t)Q_1^{-1}x(t) + \frac{1}{\lambda_1^2}\int_0^t [\|U_1x(\tau)\|^2 - \|f_{01}(x,\tau)\|^2]d\tau$$

and

$$\Gamma_2(e_w,t) = e_w^T(t)P_2e_w(t)$$

Firstly differentiating $\Gamma_2(e_w,t)$, along with the trajectories of (2.27) under (2.29) yields

$$\begin{aligned}
\dot{\Gamma}_2(e_w,t) &= e_w^T(t)sym[P_2(W + LB_0V)]e_w(t) \\
&\quad -2e_w^T P_2 LF_{01}f_{01}(x,t) \\
&\leq -\alpha_2\|e_w(t)\|^2 + \gamma_2\|x(t)\|\|e_w(t)\| \qquad (2.31)
\end{aligned}$$

Similarly, for (2.27) in the absence of $e_w(t)$, based on (2.28) after letting $P_1 = Q_1^{-1}$ we can get

$$\begin{aligned}
\dot{\Gamma}_1(x,t) &= x^T(t)[sym(P_1(A_0 + B_0K)) \\
&\quad + \lambda_1^2 P_1 F_{01}F_{01}^T P_1 + \frac{1}{\lambda_1^2}U_1^T U_1]x(t) \\
&\leq -\alpha_1\|x(t)\|^2 \qquad (2.32)
\end{aligned}$$

Furthermore, if (2.28) and (2.29) hold, there exists $\gamma_1 > 0$ depending on P_1 such that for any $x(t)$ and $e_w(t)$,

$$2x^T(t)P_1 B_0 Ve_w(t) \leq \gamma_1\|x(t)\|\|e_w(t)\| \qquad (2.33)$$

The right part of the above inequality can be regarded as a polynomial with, respect to two variables $\|x\|$ and $\|e_w\|$. Thus for all $x(t)$ and $e_w(t)$, $\dot{\Gamma}(x,e_w,t) < 0$ holds if there exist a group of parameters α_i, γ_i $(i = 1,2)$ and η_0 satisfying

$$\gamma_2^2\eta_0^2 + 2\eta_0(\gamma_1\gamma_2 - \alpha_1\alpha_2) + \gamma_1^2 < 0 \qquad (2.34)$$

The left hand of (2.34) can be considered as a polynomial with respect to η_0. There must exist an $\eta_0 > 0$ such that (2.34) holds for any given α_i and γ_i (i=1, 2) as long as (2.30) holds. $\square$

Remark 2.5 The problem is cast into solvability of a pair of LMIs with constraint (2.30) after the corresponding parameters λ and α_1 or α_2 are given properly.

Remark 2.6 From the proofs of Theorems 2.1-2.3, it is noted that the proposed approaches still have some redundant degree for the control performance. For example, in Theorem 2.1, inequality (2.14) implies that with the obtained controller, the error tracking performance and stability of the closed-loop system can also be guaranteed even if some parameter perturbations exist, where the redundance depends on the value of $min\{\sqrt{\eta_2}, \eta_3/2\sqrt{\eta_2}\}$. This will be shown in simulations.

2.6 SIMULATIONS ON A4D AIRCRAFT

2.6.1 THE CASE WITH KNOWN NONLINEARITY

The longitudinal dynamics of $A4D$ aircraft at a flight condition of 15,000-foot altitude and Mach 0.9 can be given by

$$\dot{x}(t) = A_0 x(t) + f(x(t),t) + B_0[u(t) + d(t)] \tag{2.35}$$

where $x_1(t)$ is forward velocity (ft/s), $x_2(t)$ is angle of attack (rad), $x_3(t)$ is pitching velocity (rad/s), $x_4(t)$ is pitching angle (rad), $u(t)$ is elevator deflection (deg) and

$$A_0 = \begin{bmatrix} -0.0605 & 32.37 & 0 & 32.2 \\ -0.00014 & -1.475 & 1 & 0 \\ -0.0111 & -34.72 & -2.793 & 0 \\ 0 & 0 & 1 & 0 \end{bmatrix} ; \tag{2.36}$$

$$B_0 = \begin{bmatrix} 0 \\ -0.1064 \\ -33.8 \\ 0 \end{bmatrix} ; \tag{2.37}$$

The eigenvalues of A_0 are $-2.1250 \pm j5.8510$ and -0.0039 ± 0.0896, which means the nominal system without nonlinearity is marginally stable. See References [167] and [148] for details.

Since there may be a large degree of nonlinearity and/or uncertainty in the $(3,2)$ entry of A_0, similar to Reference [167], it is supposed that

$$f(x(t),t) = F_{01} f_{01}(x(t),t) \tag{2.38}$$

where

$$F_{01} = \begin{bmatrix} 0 & 0 & 50 & 0 \end{bmatrix}^T, \ U_1 = diag\{0 \ 1 \ 0 \ 0\},$$

$$f_{01}(x(t),t) = \sin(2\pi 5t)x_2(t),$$

and

$$\|f_{01}(x(t),t)\| \le \|U_1 x(t)\|.$$

$d(t)$ is assumed to be an unknown harmonic disturbance described by (2.3) with

$$W = \begin{bmatrix} 0 & 5 \\ -5 & 0 \end{bmatrix}, \ V = \begin{bmatrix} 25 & 0 \end{bmatrix} \tag{2.39}$$

It is noted that the selection of λ is trade-off based on Remark 2.3. To avoid the high gain of controller and observer, we select $\lambda = 1$, and the initial value of the disturbance is 0.1. Suppose that the initial value of the state is taken to be

$$x(0) = \begin{bmatrix} 2 & -2 & 3 & 2 \end{bmatrix}^T.$$

When the full states can be measured, applying the approach in Theorem 2.2, we can get

$$K = \begin{bmatrix} 2.3165 & 9.9455 & 4.0004 & 13.8525 \end{bmatrix},$$

$$L = \begin{bmatrix} 0 & 0.1255 & 0.0008 & 0 \\ 0 & 0.6470 & -0.0019 & 0 \end{bmatrix}$$

and the controller is

$$u(t) = -\hat{d}(t) + Kx(t)$$

If the robust control law obtained in Reference [167] is applied, where $f(x(t),t)$ was regarded as the uncertainty, it can be shown that although the robust control law has good performance for the plant without the harmonic disturbance $d(t)$, asymptotical stability cannot be guaranteed in the presence of $d(t)$ (Figure 2.1). When the DOBC is applied, Figure 2.2 shows that asymptotical stability is achieved. Figure 2.3 illustrates the satisfactory tracking ability of our disturbance observer. Moreover, assuming all of the parameters in A_0 have $+15\%$ and -15% perturbations, Figures 2.4 and 2.5 show that the satisfying robustness performance can also be guaranteed, since with these solutions, corresponding to (2.24), $\alpha = 0.1366$, which means even if some perturbations occur, (2.24) (or $\Pi_2 < 0$) may still hold (see also Remark 2.6). So does $\Pi_2 < 0$.

It should be pointed out that if the frequency is perturbed (W is perturbed in (2.3)), the DOBC approach will be unavailable because the disturbances cannot be rejected accurately in this case. Concretely, in this situation, $\hat{d}(t)$ will converge to another disturbance resulting from the perturbed disturbance dynamics. Figure 2.6 shows that only boundedness rather than asymptotical stability can be guaranteed if frequency turns to be 5.05. For an unstable or unbounded disturbance signal described by (2.3) with

$$W = \begin{bmatrix} 0.1 & 1 \\ 0 & 0.1 \end{bmatrix}, V = \begin{bmatrix} 2 & 0 \end{bmatrix}$$

Figure 2.7 demonstrates that, by using Theorem 2.2, stability can be guaranteed in the local sense (see Reference [234]), where the solid lines represent the responses of the state while the dashed one represents the response of the estimation error.

2.6.2 THE CASE WITH UNKNOWN NONLINEARITY

To compare with the previous works, we also consider the longitudinal dynamics of A4D aircraft described as (2.35). Similarly, $d(t)$ is assumed to be an unknown harmonic disturbance described by (2.3) and (2.39). Different from the above subsection, we suppose that $f(x(t),t)$ is unknown but satisfies (2.37) and (2.38). In simulation, we assume

$$f_{01}(x(t),t) = \begin{bmatrix} 0 & r(t)x_2(t) & 0 & 0 \end{bmatrix}^T$$

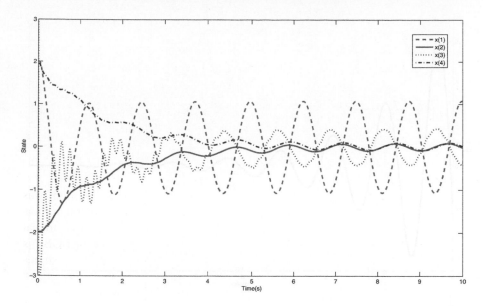

FIGURE 2.1   System performance using robust control without system disturbances.

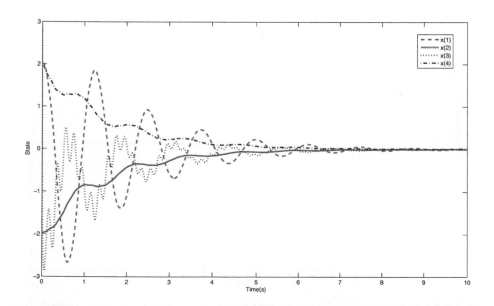

FIGURE 2.2   System performance using DOBC with system disturbances.

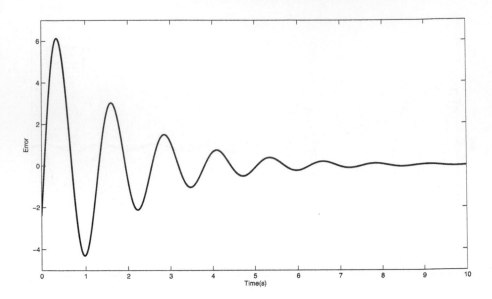

FIGURE 2.3   Estimation error for system disturbances using DOBC.

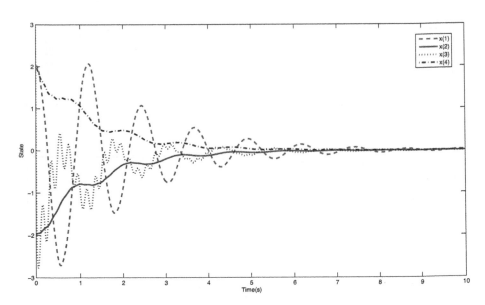

FIGURE 2.4   System performance using DOBC with system disturbances and +15% parametric perturbations.

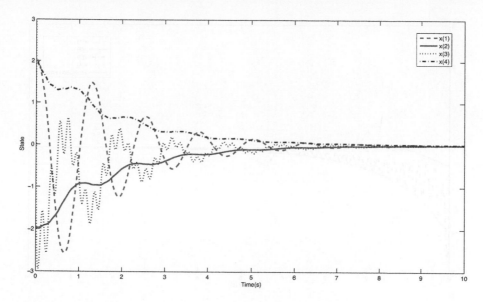

FIGURE 2.5   System performance using DOBC with system disturbances and -15% parametric perturbations.

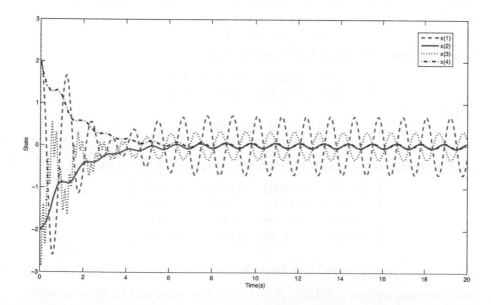

FIGURE 2.6   System performance using DOBC with perturbed system disturbances.

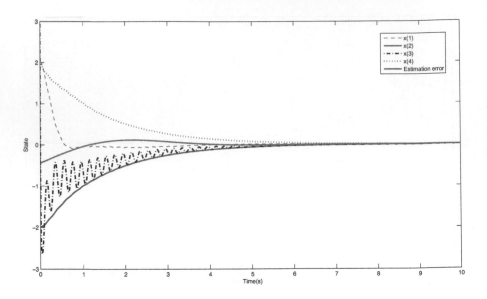

FIGURE 2.7   System performance using DOBC with unbounded system disturbances.

where $r(t)$ is assumed to be a random input with an upper bound 50. It is noted that the case cannot be treated by the results in Sections 2.3 and 2.4. Suppose the initial value is

$$x(0) = \begin{bmatrix} 2 & -2 & 1 & -2 \end{bmatrix}^T$$

Based on Theorem 2.3, from (2.29) it can be obtained that

$$R_2 = \begin{bmatrix} 0 & 25.2520 & 0.3968 & 0 \\ 0 & 257.8980 & -0.8118 & 0 \end{bmatrix},$$

$$L = \begin{bmatrix} 0 & 0.1255 & 0.0008 & 0 \\ 0 & 0.6470 & -0.0019 & 0 \end{bmatrix}$$

and correspondingly $\alpha_2 = 398.3653$, $\gamma_2 = 90.3631$. From (2.28) we can get

$$P_1 = \begin{bmatrix} 0.0000 & -0.0005 & 0.0000 & 0.0008 \\ -0.0005 & 0.1944 & -0.0003 & -0.1389 \\ 0.0000 & -0.0003 & 0.0000 & 0.0006 \\ 0.0008 & -0.1389 & 0.0006 & 0.1719 \end{bmatrix},$$

$$K = \begin{bmatrix} 0.1587 & 46.4563 & 0.1597 & 26.7096 \end{bmatrix}$$

and correspondingly $\alpha_1 = 0.1599$, $\gamma_1 = 0.5836$. It is noted that (2.30) is satisfied. Under the proposed robust DOBC law, Figure 2.8 shows the stability of the closed-loop system and Figure 2.9 shows the convergence to track the external disturbance. If all of the coefficients of A_0 have $+30\%$ parametric perturbation, Figure 2.10 demonstrates that the performance robustness can also be guaranteed.

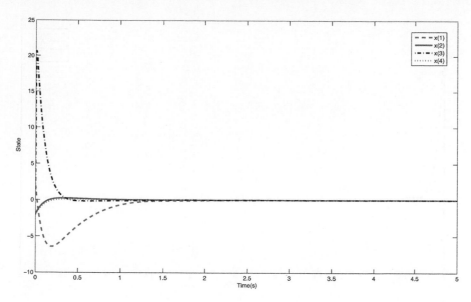

FIGURE 2.8 System performance using DOBC with system disturbances and unknown nonlinearity.

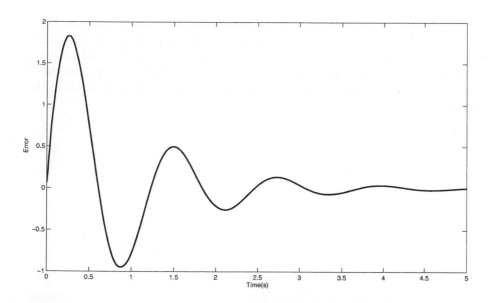

FIGURE 2.9 Estimation errors for disturbances with unknown system nonlinearity.

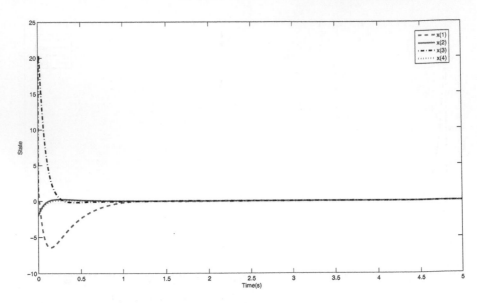

FIGURE 2.10   System performance using DOBC with system disturbances, unknown non-linearity and parametric perturbations.

2.7   CONCLUSION

In this chapter, new DOBC approaches are investigated for a class of MIMO nonlinear systems with disturbances. Feasible design procedures are proposed under different conditions to estimate and reject disturbance for plants with known and unknown nonlinearity, respectively. Based on the estimation of disturbances, composite control laws can guarantee that composite closed-loop systems are globally stable in the presence of disturbances.

3 Composite DOBC and H_∞ Control for Continuous Models

3.1 INTRODUCTION

To improve anti-disturbance ability, disturbance observer based control (DOBC) strategies have been investigated and applied in many control areas [32, 85, 89, 108, 110, 121, 122, 133, 137, 197, 237]. DOBC has its roots in many mechanical applications in the last two decades, in particular for linear systems [108, 110, 121]. Attempts have been made to establish theoretical justification of these DOBC applications and extend DOBC from linear systems to nonlinear systems [32, 85], similar to nonlinear regulation theory [111, 234]. In [32], only single-input-single-output (SISO) nonlinear systems with well-defined disturbance relative degree were studied, and the disturbances were limited to be constant or harmonic signals. In [85], the DOBC approaches for a class of multiple-input-multiple-output (MIMO) nonlinear systems have been considered, where the nonlinear dynamics can be described by known and unknown nonlinear functions, respectively, and the disturbances were represented by a linear exogenous system. This extended the assumptions of the disturbances, which are restricted to being constant, harmonic or neutral stable [32, 111, 234]. However, it has been reported that when a disturbance has perturbations, the proposed approaches in DOBC (e.g., in [32, 85]) are unsatisfactory, which has been verified by the simulations in [85].

In some cases, the perturbations and unmodeling errors can be merged into the disturbance, which can be supposed to be bounded by the appropriate norms (e.g., H_∞ control). Apart from other bounded noise (e.g., stochastic white noise), it is known that some disturbance (such as harmonic disturbance or periodic noise) can be described by an exogenous system as investigated in [32, 85, 111]. However, uncertainties cannot be considered in such an exogenous model for disturbance, which restricts the application of the results in [32, 85] (also in all other literature where the exogenous system is assumed to have a precise model).

The objective of this chapter is to investigate composite control for complex continuous models subject to disturbances, where the disturbance includes two parts. One part is bounded vector in an H_2-norm context. The other part is supposed to be generated by an exogenous system, which is not required to be neutral stable as in the output regulation theory. Notably, the exosystem is also supposed to have modeling perturbation. A novel composite DOBC and H_∞ control scheme is presented so that the disturbance with the exogenous system can be estimated and compensated for, and disturbance with the bounded norm can be attenuated using H_∞ control law.

A reduced-order observer based on regional pole placement and D-stability theory is structured for the estimation of the disturbance, where the modeling errors and perturbations in the exogenous subsystem can be formulated into the disturbance.

3.2 FORMULATION OF THE PROBLEM

The following MIMO continuous system with two types of disturbances and nonlinear dynamics is described as

$$\dot{x}(t) = G_0 x(t) + F_{01} f_{01}(x(t), t) + H_0[u(t) + d_0(t)] + H_1 d_1(t) \qquad (3.1)$$

where $x(t) \in R^n$, $u(t) \in R^m (m < n)$ is state and control input, respectively. $G_0 \in R^{n \times n}$, and $H_0 \in R^{n \times m}$, $H_1 \in R^{n \times p}$ are the coefficient matrices. $F_{01} \in R^{n \times q}$ are the corresponding weighting matrices. $f_{01} \in R^q$ are nonlinear vector functions which are supposed to satisfy bounded conditions described as Assumption 3.2. $d_0(t) \in R^m$ is supposed to be described by an exogenous system in Assumption 3.1, which can represent the constant and the harmonic noises. $d_1(t) \in R^p$ is the external disturbance in the H_2-norm.

Assumption 3.1 *The disturbance $d_0(t)$ in the control input path can be formulated by the following exogenous system*

$$\begin{cases} \dot{w}(t) &= Ww(t) + H_2 \delta(t) \\ d_0(t) &= Vw(t) \end{cases} \qquad (3.2)$$

where $W \in R^{r \times r}$, $H_2 \in R^{r \times l}$ and $V \in R^{m \times r}$ are proper known matrices. $\delta(t) \in R^l$ is the additional disturbance which results from the perturbations and uncertainties in the exogenous system. It is also supposed that $\delta(t)$ is bounded in the H_2-norm.

In many cases, system disturbance can be described as a dynamical system with unknown parameters and initial conditions [32, 85, 111, 234]. However, uncertainties have been included in such an exogenous model for disturbance, which restricts the application of the results in [32, 85] (also in all other literature where the exogenous system is assumed to have a precise model). In this chapter, besides the other bounded noises, $d_0(t)$ can also be used to include the unmodeling error and system perturbations. Similarly to [32, 81], the following assumption is required to describe the nonlinear dynamics.

Assumption 3.2 *The nonlinear function satisfies*

$$\|f_{01}(x_1(t), t) - f_{01}(x_2(t), t)\| \leq \|U_1(x_1(t) - x_2(t))\| \qquad (3.3)$$

where U_1 is a given constant weighting matrix. The following assumption is a necessary condition for the DOBC formulation.

Assumption 3.3 *(G_0, H_0) is controllable and $(W, H_0 V)$ is observable.*

Suppose that the system state is available. If $d_0(t)$ is considered as a part of the augmented state, then a reduced-order observer is needed. In this chapter, we construct the reduced-order observer for $d_0(t)$ for the case with known nonlinearity and unknown nonlinearity, respectively. And then we design a composite controller with the disturbance observer and a H_∞ controller so that the disturbance can be rejected and attenuated, simultaneously; the stability of the resulting composite system can also be guaranteed.

3.3 COMPOSITE DOBC AND H_∞ CONTROL FOR THE CASE WITH KNOWN NONLINEARITY

In this section, we suppose that f_{01} are given and Assumptions 3.1-3.3 hold. When all states of the system are available, it is unnecessary to estimate the states, and only the estimation of the disturbance needs to be included.

3.3.1 DISTURBANCE OBSERVER

In this section, the disturbance observer is constructed as

$$
\begin{cases}
\hat{d}_0(t) &= V\hat{w}(t) \\
\hat{w}(t) &= v(t) - Lx(t) \\
\dot{v}(t) &= (W + LH_0V)[v(t) - Lx(t)] + L[G_0x(t) \\
&\quad + F_{01}f_{01}(x(t),t) + H_0u(t)]
\end{cases}
\tag{3.4}
$$

where $\hat{w}(t)$ is the estimation of $w(t)$, and $v(t)$ is the auxiliary vector as the state of the observer. The estimation error is denoted as $e_w(t) = w(t) - \hat{w}(t)$. Based on (3.1), (3.2) and (3.4), it is shown that the error dynamics satisfies

$$
\dot{e}_w(t) = (W + LH_0V)e_w(t) + H_2\delta(t) + LH_1d_1(t)
\tag{3.5}
$$

The objective of disturbance rejection can be achieved by designing the observer gain such that (3.5) satisfies the desired stability and robustness performance. The structure of the controller is formulated as

$$
u(t) = -\hat{d}_0(t) + Kx(t)
\tag{3.6}
$$

Combining (3.6) with (3.1), the closed-loop system is described as

$$
\dot{x}(t) = (G_0 + H_0K)x(t) + F_{01}f_{01}(x(t),t) + H_0Ve_w(t) + H_1d_1(t)
\tag{3.7}
$$

Thus, the composite system combined (3.5) with (3.7) yields

$$
\begin{bmatrix} \dot{x}(t) \\ \dot{e}_w(t) \end{bmatrix} = \begin{bmatrix} G_0 + H_0K & H_0V \\ 0 & W + LH_0V \end{bmatrix} \begin{bmatrix} x(t) \\ e_w(t) \end{bmatrix} + \begin{bmatrix} F_{01} \\ 0 \end{bmatrix} f_{01}(x(t),t)
$$
$$
+ \begin{bmatrix} H_1 & 0 \\ LH_1 & H_2 \end{bmatrix} \begin{bmatrix} d_1(t) \\ \delta(t) \end{bmatrix}
$$

And the concerned system can be formulated by

$$
\begin{cases}
\dot{\bar{x}}(t) &= G\bar{x}(t) + Ff(\bar{x}(t),t) + Hd(t) \\
z(t) &= C\bar{x}(t)
\end{cases}
\tag{3.8}
$$

where

$$
\bar{x}(t) = \begin{bmatrix} x(t) \\ e_w(t) \end{bmatrix}; \quad
G = \begin{bmatrix} G_0 + H_0 K & H_0 V \\ 0 & W + LH_0 V \end{bmatrix};
$$

$$
F = \begin{bmatrix} F_{01} \\ 0 \end{bmatrix}; \quad f(\bar{x}(t),t) = f_{01}(x(t),t);
$$

$$
H = \begin{bmatrix} H_1 & 0 \\ LH_1 & H_2 \end{bmatrix}; \quad d(t) = \begin{bmatrix} d_1(t) \\ \delta(t) \end{bmatrix}
$$

In (3.8), the reference output is set to be $z(t) = C\bar{x}(t)$, where $C = \begin{bmatrix} C_1 & C_2 \end{bmatrix}$ is weighting matrices to adjust the system performance. For $f(\bar{x}(t),t)$, it can be seen that

$$
\|f(\bar{x}(t),t)\| \le \|U\bar{x}(t)\|
$$

where

$$
U = \begin{bmatrix} U_1 & 0 \\ 0 & 0 \end{bmatrix}
$$

and U_1 is given in (3.3).

With the above formulations, it can be seen that the composite system includes two subsystems. One is the error dynamical system for the estimation of the disturbance governed by the exogenous system. The other results from the original system by using the disturbance rejection term in control input and merging two types of disturbance into an augmented one. The problem considered in this work is stated as follows: to design an observer to estimate the first type of disturbance, and compute a composite H_∞ controller such that (3.8) is stable and satisfies $\|z(t)\|_2 \le \gamma \|d(t)\|_2$, where γ is a given positive constant for the level of disturbance attenuation. The new control scheme combines the classical DOBC with H_∞ control, which can be called the "DOBC plus H_∞ control," and abbreviated as DOBPH_∞C.

3.3.2 COMPOSITE DOBC AND H_∞ CONTROL

In this section, the main task is to design L and K such that (3.8) is asymptotically stable and satisfies the performance of disturbance attenuation. For the convenience of research, the following lemma is presented.

Lemma 3.1

For the following system

$$\begin{cases} \dot{x}(t) &= Gx(t) + Ff(x(t),t) + Hd(t) \\ y(t) &= Cx(t) \end{cases} \tag{3.9}$$

for given parameters $\lambda > 0$, $\gamma > 0$, if there exist $P > 0$, satisfying

$$\begin{bmatrix} PG + G^T P + \frac{1}{\lambda^2} U_*^T U_* & PF & PH & C^T \\ F^T P & -\frac{1}{\lambda^2} I & 0 & 0 \\ H^T P & 0 & -\gamma^2 I & 0 \\ C & 0 & 0 & -I \end{bmatrix} < 0 \tag{3.10}$$

then (3.9) is robustly asymptotically stable in the absence of the disturbance $d(t)$, and satisfies $\|z(t)\|_2 < \gamma \|d(t)\|_2$.
Where

$$\|f(x_1(t),t) - f(x_2(t),t)\| \le \|U_*(x_1(t) - x_2(t))\|$$

and U_* are given constant weighting matrices. ∎

Proof. For the system (3.9)

$$\begin{cases} \dot{x}(t) &= Gx(t) + Ff(x(t),t) + Hd(t) \\ y(t) &= Cx(t) \end{cases}$$

Consider the following Lyapunov function

$$V(t) = x^T(t)Px(t) + \frac{1}{\lambda^2} \int_0^t [\|U_* x(\tau)\|^2 - \|f(x(\tau),\tau)\|^2] d\tau \tag{3.11}$$

Along with the trajectories of (3.9) in the absence of the disturbance $d(t)$, it can be shown that

$$\begin{aligned} \dot{V}(t) &= \dot{x}^T(t)Px(t) + x^T(t)P\dot{x}(t) + \frac{1}{\lambda^2}[\|U_* x(t)\|^2 - \|f(x(t),t)\|^2] \\ &= x^T(t)PGx(t) + x^T(t)PFf(x(t),t) + f^T(x(t),t)F^T Px(t) \\ &\quad + x^T(t)PG^T x(t) + \frac{1}{\lambda^2}x^T(t)U_*^T U_* x(t) - \frac{1}{\lambda^2}f^T(x(t),t)f(x(t),t) \\ &= \begin{bmatrix} x(t) \\ f(x(t),t) \end{bmatrix}^T \begin{bmatrix} PG + G^T P + \frac{1}{\lambda^2}U_*^T U_* & PF \\ F^T P & -\frac{1}{\lambda^2}I \end{bmatrix} \begin{bmatrix} x(t) \\ f(x(t),t) \end{bmatrix} \\ &= \bar{x}^T(t)Q_0\bar{x}(t) \end{aligned} \tag{3.12}$$

where

$$\bar{x}(t) = \left[\begin{array}{c} x(t) \\ f(x(t),t) \end{array} \right],$$

$$Q_0 = \left[\begin{array}{cc} PG + G^T P + \frac{1}{\lambda^2} U_*^T U_* & PF \\ F^T P & -\frac{1}{\lambda^2} I \end{array} \right]$$

Based on the Lyapunov theory, it is noted that the system (3.9) is asymptotically stable in the absence of $d(t)$ if $Q_0 < 0$ holds.

The next step is to focus on the condition of disturbance attenuation. The following auxiliary function (known as the storage function) is considered

$$J(x(t)) = V(t) + \int_0^t [\|z(t)\|^2 - \gamma^2 \|d(t)\|^2] dt \tag{3.13}$$

which satisfies that $J(x(t)) = \int_0^t S(\tau) d\tau$ with the zero initial condition, where $V(t)$ is denoted as in (3.11),

$$S(t) = \|z(t)\|^2 - \gamma^2 \|d(t)\|^2 + \dot{V}(t),$$

then it can be verified that

$$
\begin{aligned}
S(t) &= x^T(t)C^T Cx(t) - \gamma^2 d^T(t)d(t) + \dot{x}^T(t)Px(t) + x^T(t)P\dot{x}(t) \\
&\quad + \frac{1}{\lambda^2}[\|U_* x(t)\|^2 - \|f(x(t),t)\|^2] \\
&= x^T(t)C^T Cx(t) + x^T(t)P[Gx(t) + Ff(x(t),t) + Hd(t)] \\
&\quad + [x^T(t)G^T + f^T(x(t),t)F + d^T(t)H^T]Px(t) - \gamma^2 d^T(t)d(t) \\
&\quad + \frac{1}{\lambda^2}x^T(t)U_*^T U_* x(t) - \frac{1}{\lambda^2}f^T(x(t),t)f(x(t),t) \\
&= \bar{x}^T(t)Q_1 \bar{x}(t) \tag{3.14}
\end{aligned}
$$

where

$$\bar{x}(t) = \left[\begin{array}{c} x(t) \\ f(x(t),t) \\ d(t) \end{array} \right],$$

$$Q_1 = \left[\begin{array}{ccc} PG + G^T P + \frac{1}{\lambda^2} U_*^T U_* + C^T C & PF & PH \\ F^T P & -\frac{1}{\lambda^2} I & 0 \\ H^T P & 0 & -\gamma^2 I \end{array} \right]$$

It can be seen that $Q_1 < 0$ if and only if $Q_2 < 0$ based on the Schur complement formula, where $Q_2 < 0$ is denoted as in (3.10). Hence, (3.10) implies that $Q_1 < 0$

and $S(t) < 0$. It can be easily obtained that $J(t) < 0$ if $S(t) < 0$ hold, which further leads to $\|z(t)\|_2^2 < \gamma^2 \|d(t)\|_2^2$ holds, which is equivalent to $\|z(t)\|_2 < \gamma \|d(t)\|_2$. On the other hand, it can be verified that $Q_2 < 0$ also implies that $Q_0 < 0$ by deleting the 3rd row and 3rd column of $Q_1 < 0$. Thus, the system (3.9) is asymptotically stable in the absence of $d(t)$ if $Q_2 < 0$ holds. The proof is completed. □

Based on Lemma 3.1, we have the following theorem.

Theorem 3.1

For given parameters $\lambda > 0$, $\gamma > 0$ and $\Theta > 0$, if there exist $Q_1 > 0$, $P_2 > 0$ and R_1, R_2 satisfying

$$\begin{bmatrix} M_1 & F_{01} & H_1 & 0 & Q_1 C^T & Q_1 U_1^T & H_0 V \\ * & -\frac{1}{\lambda^2}I & 0 & 0 & 0 & 0 & 0 \\ * & * & -\gamma^2 I & 0 & 0 & 0 & H_1^T R_2^T \\ * & * & * & -\gamma^2 I & 0 & 0 & H_2^T P_2^T \\ * & * & * & * & -I & 0 & C_2 \\ * & * & * & * & * & -\lambda^2 I & 0 \\ * & * & * & * & * & * & M_2 \end{bmatrix} < 0 \qquad (3.15)$$

$$P_2 W + W^T P_2 + R_2 H_0 V + V^T H_0^T R_2^T + P_2 \Theta < 0 \qquad (3.16)$$

where

$$M_1 = G_0 Q_1 + Q_1^T G_0^T + H_0 R_1 + R_1^T H_0^T,$$

$$M_2 = P_2 W + W^T P_2 + R_2 H_0 V + V^T H_0^T R_2^T$$

then composite system (3.8) under DOBC law (3.6) with gain $K = R_1 Q_1^{-1}$ and observer (3.4) with gain $L = P_2^{-1} R_2$ is robustly asymptotically stable in the absence of the disturbance $d(t)$, and satisfies $\|z(t)\|_2 < \gamma \|d(t)\|_2$. ∎

Proof. For system (3.8), we denote

$$P = \begin{bmatrix} P_1 & 0 \\ 0 & P_2 \end{bmatrix} = \begin{bmatrix} Q_1^{-1} & 0 \\ 0 & P_2 \end{bmatrix} > 0 \qquad (3.17)$$

Applying Lemma 3.1 to (3.8), it can be verified that

$$\Omega_1 = \begin{bmatrix} \Pi_1 + \frac{1}{\lambda^2}U_1^T U_1 & P_1 H_0 V & P_1 F_{01} & P_1 H_1 & 0 & C_1^T \\ * & \Pi_2 & 0 & P_2 L H_1 & P_2 H_2 & C_2^T \\ * & * & -\frac{1}{\lambda^2}I & 0 & 0 & 0 \\ * & * & * & -\gamma^2 I & 0 & 0 \\ * & * & * & * & -\gamma^2 I & 0 \\ * & * & * & * & * & -I \end{bmatrix} < 0 \qquad (3.18)$$

then closed-loop system (3.8) is robustly asymptotically stable in the absence of the disturbance $d(t)$, and satisfies $||z(t)||_2 < \gamma ||d(t)||_2$ if $\Omega_1 < 0$ holds. Where

$$\Pi_1 = P_1(G_0 + H_0K) + (G_0 + H_0K)^T P_1,$$

$$\Pi_2 = P_2(W + LH_0V) + (W + LH_0V)^T P_2$$

Based on Schur complement, $\Omega_1 < 0$ is equivalent to $\Omega_2 < 0$, where

$$\Omega_2 = \begin{bmatrix} \Pi_1 & P_1H_0V & P_1F_{01} & P_1H_1 & 0 & C_1^T & U_1^T \\ * & \Pi_2 & 0 & P_2LH_1 & P_2H_2 & C_2^T & 0 \\ * & * & -\frac{1}{\lambda^2}I & 0 & 0 & 0 & 0 \\ * & * & * & -\gamma^2I & 0 & 0 & 0 \\ * & * & * & * & -\gamma^2I & 0 & 0 \\ * & * & * & * & * & -I & 0 \\ * & * & * & * & * & * & -\lambda^2I \end{bmatrix} < 0 \quad (3.19)$$

Exchanging of rows and columns, $\Omega_2 < 0$ is equivalent to $\Omega_3 < 0$, where

$$\Omega_3 = \begin{bmatrix} \Pi_1 & P_1F_{01} & P_1H_1 & 0 & C_1^T & U_1^T & P_1H_0V \\ * & -\frac{1}{\lambda^2}I & 0 & 0 & 0 & 0 & 0 \\ * & * & -\gamma^2I & 0 & 0 & 0 & (P_2LH_1)^T \\ * & * & * & -\gamma^2I & 0 & 0 & (P_2H_2)^T \\ * & * & * & * & -I & 0 & C_2 \\ * & * & * & * & * & -\lambda^2I & 0 \\ * & * & * & * & * & * & \Pi_2 \end{bmatrix} < 0 \quad (3.20)$$

Ω_3 is pre-multiplied and post-multiplied simultaneously by $diag\{Q_1, I, I, I, I, I, I\}$, then is equivalent to Ω_4, where

$$\Omega_4 = \begin{bmatrix} M_1 & F_{01} & H_1 & 0 & Q_1C_1^T & Q_1U^T & H_0V \\ * & -\frac{1}{\lambda^2}I & 0 & 0 & 0 & 0 & 0 \\ * & * & -\gamma^2I & 0 & 0 & 0 & H_1^T R_2^T \\ * & * & * & -\gamma^2I & 0 & 0 & H_2^T P_2^T \\ * & * & * & * & -I & 0 & C_2 \\ * & * & * & * & * & -\lambda^2I & 0 \\ * & * & * & * & * & * & M_2 \end{bmatrix} < 0 \quad (3.21)$$

Thus (3.15) can be obtained.

The next step, according to regional pole placement and D-stability theory [225], choose regional pole $\Theta > 0$ for (3.5), then exist $X > 0$, it can be verified that

$$WX + X^TW^T + LH_0VX + X^TV^TH_0^TL^T + \Theta X < 0$$

Then the pole will be assigned following LMI region D.

$$D = \{s \in C : \Theta + sM + \bar{s}M^T < 0\}$$

(3.16) can be obtained by defining $X = P_2^{-1}$, $L = P_2 R_2$. Then by associating (3.15) with (3.16) and selecting $K = R_1 Q_1^{-1}$, observer gain $L = P_2^{-1} R_2$, the composite system (3.8) is robustly asymptotically stable and satisfying $||z(t)||_2 < \gamma ||d(t)||_2$. The proof is completed. $\qquad \square$

Remark 3.1 The observer gain $L = P_2^{-1} R_2$ is obtained based on regional pole placement and D-stability theory. And DOBPH_∞C can be mixed by pole placement and D-stability theory using LMI, which allows a better system performance for the closed-loop behavior.

3.4 COMPOSITE DOBC AND H_∞ CONTROL FOR THE CASE WITH UNKNOWN NONLINEARITY

In this section, we suppose the Assumptions 3.1-3.3 hold, but nonlinear functions f_{01} are unknown. Different from Section 3.3, f_{01} is unavailable in observer design.

3.4.1 DISTURBANCE OBSERVER

In this section, construct the disturbance observer as

$$\begin{cases} \dot{\hat{d}}_0(t) &= V\hat{w}(t) \\ \hat{w}(t) &= v(t) - Lx(t) \\ \dot{v}(t) &= (W + LH_0 V)(v(t) - Lx(t)) + L(G_0 x(t) + H_0 u(t)) \end{cases} \tag{3.22}$$

Compared with (3.5), the estimation error $e_w(t) = w(t) - \hat{w}(t)$ satisfies

$$\dot{e}_w(t) = (W + LH_0 V)e_w(t) + LF_{01} f_{01} + H_2 \delta(t) + LH_1 d_1(t) \tag{3.23}$$

The controller is formulated as

$$u(t) = -\hat{d}_0(t) + Kx(t) \tag{3.24}$$

Combining (3.24) with (3.1), the closed-loop system is described as

$$\dot{x}(t) = (G_0 + H_0 K)x(t) + F_{01} f_{01}(x(t), t) + H_0 V e_w(t) + H_1 d_1(t) \tag{3.25}$$

Thus, the composite system combined (3.23) with (3.25) is given by

$$\begin{bmatrix} \dot{x}(t) \\ \dot{e}_w(t) \end{bmatrix} = \begin{bmatrix} G_0 + H_0 K & H_0 V \\ 0 & W + LH_0 V \end{bmatrix} \begin{bmatrix} x(t) \\ e_w(t) \end{bmatrix} + \begin{bmatrix} F_{01} \\ LF_{01} \end{bmatrix} f_{01}(x(t), t)$$

$$+ \begin{bmatrix} H_1 & 0 \\ LH_1 & H_2 \end{bmatrix} \begin{bmatrix} d_1(t) \\ \delta(t) \end{bmatrix}$$

And the concerned system can be formulated by

$$\begin{cases} \dot{\bar{x}}(t) &= G\bar{x}(t) + Ff(\bar{x}(t), t) + Hd(t) \\ z(t) &= C\bar{x}(t) \end{cases} \tag{3.26}$$

where

$$\bar{x}(t) = \begin{bmatrix} x(t) \\ e_w(t) \end{bmatrix}; \quad G = \begin{bmatrix} G_0 + H_0K & H_0V \\ 0 & W + LH_0V \end{bmatrix};$$

$$F = \begin{bmatrix} F_{01} \\ LF_{01} \end{bmatrix}; \quad f(\bar{x}(t),t) = f_{01}(x(t),t);$$

$$H = \begin{bmatrix} H_1 & 0 \\ LH_1 & H_2 \end{bmatrix}; \quad d(t) = \begin{bmatrix} d_1(t) \\ \delta(t) \end{bmatrix}$$

In (3.26), $z(t) = C\bar{x}(t)$ is denoted as the reference output, where $C = \begin{bmatrix} C_1 & C_2 \end{bmatrix}$ is the weighting matrices.

3.4.2 COMPOSITE DOBC AND H_∞ CONTROL

Similar to Section 3.2, in this section our aim is to design L and K in order that (3.26) is asymptotically stable and satisfies the performance of disturbance attenuation. So applying Lemma 3.1 to (3.26), we obtain

---

Theorem 3.2

For given parameters $\lambda > 0$, $\gamma > 0$ and $\Theta > 0$, if there exist $Q_1 > 0$, $P_2 > 0$ and R_1, R_2 satisfying

$$\begin{bmatrix} M_1 & F_{01} & H_1 & 0 & Q_1C^T & Q_1U_1^T & H_0V \\ * & -\frac{1}{\lambda^2}I & 0 & 0 & 0 & 0 & F_{01}^T R_2^T \\ * & * & -\gamma^2 I & 0 & 0 & 0 & H_1^T R_2^T \\ * & * & * & -\gamma^2 I & 0 & 0 & H_2^T P_2^T \\ * & * & * & * & -I & 0 & C_2 \\ * & * & * & * & * & -\lambda^2 I & 0 \\ * & * & * & * & * & * & M_2 \end{bmatrix} < 0 \qquad (3.27)$$

$$P_2W + W^T P_2 + R_2H_0V + V^T H_0^T R_2^T + P_2\Theta < 0 \qquad (3.28)$$

where

$$M_1 = G_0Q_1 + Q_1^T G_0^T + H_0R_1 + R_1^T H_0^T,$$

$$M_2 = P_2W + W^T P_2 + R_2H_0V + V^T H_0^T R_2^T$$

then by selecting $K = R_1Q_1^{-1}$ and $L = P_2^{-1}R_2$, the composite system (3.26) is asymptotically stable in the absence of the disturbance $d(t)$, and satisfying $\|z(t)\|_2 < \gamma\|d(t)\|_2$. ∎

Proof. Noting the difference of F between (3.8) and (3.26), the proof procedure can be given similarly to that of the proof for Theorem 3.1. □

3.5 SIMULATION EXAMPLE

In [85], an A4D aircraft model in a continuous-time context has been used to show that the pure DOBC approach provides more robust performance against disturbance and uncertainties than some previous results. However, it was also pointed out that when the exogenous system has a perturbation, the result is unsatisfactory. In this section, to compare with the work in [85], we also consider the longitudinal dynamics of A4D aircraft described as [85]. Suppose an aircraft model described by (3.1) with the following coefficient matrices.

$$G_0 = \begin{bmatrix} 0.065 & 32.37 & 0 & 32.2 \\ -0.00014 & -1.475 & 1 & 0 \\ -0.0111 & -34.72 & -2.793 & 0 \\ 0 & 0 & 1 & 0 \end{bmatrix};$$

$$H_0 = \begin{bmatrix} 0 \\ -0.1064 \\ -33.8 \\ 0 \end{bmatrix}; H_1 = \begin{bmatrix} 0.01 \\ 0 \\ -0.038 \\ 0.01 \end{bmatrix}; H_2 = \begin{bmatrix} 0.022 \\ 0.015 \end{bmatrix}$$

The reference output is set to be

$$z(t) = C\bar{x}(t)$$

where

$$C = \begin{bmatrix} C_1 & C_2 \end{bmatrix}$$

and

$$C_1 = \begin{bmatrix} 1 & 0 & 0 & 0 \end{bmatrix}; C_2 = \begin{bmatrix} 0 & 0 \end{bmatrix}$$

The exogenous system matrices for $d_0(t)$ described by (3.2) are given by

$$W = \begin{bmatrix} 0 & 5 \\ -5 & 0 \end{bmatrix}; V = \begin{bmatrix} 25 & 0 \end{bmatrix}; H_2 = \begin{bmatrix} 0.1 \\ 0.1 \end{bmatrix} \tag{3.29}$$

In simulation, we select $\delta(t)$ as the random signal with upper 2-norm bound 1 and $d_1(t)$ as continuous wind gust model. The mathematical representation is

$$V_{wind} = \begin{cases} 0, & x < 0; \\ \frac{V_m}{2}(1 - \cos(\frac{\pi x}{d_m})), & 0 \le x \le d_m; \\ V_m, & x > d_m \end{cases}$$

where V_m is the gust amplitude, d_m is the gust length, x is the distance traveled, and V_{wind} is the resultant wind velocity in the body axis frame.

3.5.1 THE CASE WITH KNOWN NONLINEARITY

Suppose that the nonlinear dynamics is denoted by

$$f_{01}(x(t),t) = sin(2\pi * 5t)x_2(t), \|f_{01}(x(t),t)\| \leq \|U_1 x(t)\|$$

where $U_1 = diag\{0,1,0,0\}$. To avoid the high gain of controller and observer, we select $\lambda = 1$. The initial value of the state is taken to be

$$x(0) = \begin{bmatrix} 2 & -2 & 3 & 2 \end{bmatrix}^T$$

Based on Theorem 3.1, it can be solved that

$$K = \begin{bmatrix} 0.8833 & 4.0620 & 0.8808 & 5.3690 \end{bmatrix};$$

$$L = \begin{bmatrix} 3.0809 & 3.7359 & 0.0161 & 3.0811 \\ 2.5851 & 3.1349 & 0.0135 & 2.5853 \end{bmatrix}$$

Figures 3.1 and 3.2 demonstrate the system performance using the DOBC scheme in [85] and the proposed DOBPH_∞C schemes, respectively. Figures 3.3 and 3.4 illustrate the estimation error for system disturbances with the DOBC in [85] and the DOBPH_∞C, respectively.

For the case with known nonlinearity, the effectiveness of the proposed DOBPH_∞C scheme can be seen in Figure 3.2. The simulation results show that although there are exogenous wind disturbances in the system, the disturbance rejection performance is improved and satisfactory system responses can be achieved with the proposed DOBPH_∞C compared with only the DOBC scheme in Figure 3.1. Figure 3.4 shows the tracking ability of our disturbance observer based on regional pole placement and D-stability theory is more satisfying than the pure DOBC scheme in Figure 3.3.

3.5.2 THE CASE WITH UNKNOWN NONLINEARITY

In comparison with the work in [85], we also consider the longitudinal dynamics of A4D aircraft described as [85]. Similarly, $d_0(t)$ is assumed to be an unknown harmonic disturbance described by (3.2) and (3.29). Different from Section 3.1, we suppose that $f_{01}(x(t),t)$ is unknown. In simulation, we assume

$$f_{01}(x(t),t) = r(t)x_2(t)$$

where $r(t)$ is assumed to be a random input with an upper bound 1. It is noted that the case cannot be treated by considering the results in Section 3.3. Suppose the initial value is $x(0) = \begin{bmatrix} 2 & -2 & 3 & 2 \end{bmatrix}^T$. Based on Theorem 3.2, it can be obtained that

$$K = \begin{bmatrix} 0.7193 & 3.2166 & 0.7447 & 4.5983 \end{bmatrix};$$

$$L = \begin{bmatrix} -0.0022 & 1.6647 & 0.0001 & -0.0022 \\ -0.0009 & 0.5493 & 0.0000 & -0.0009 \end{bmatrix}$$

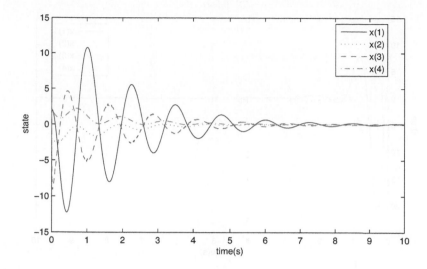

FIGURE 3.1   System performance using DOBC with disturbances for known nonlinearity.

Figures 3.5 and 3.6 demonstrate the system performance with system disturbances for the case with unknown nonlinearity using the DOBC scheme in [85] and the DOBPH_∞C schemes, respectively. Figures 3.7 and 3.8 illustrate the estimation error for system disturbances for the case with unknown nonlinearity with the DOBC in [85] and the DOBPH_∞C, respectively.

For the case with unknown nonlinearity, Figure 3.6 shows that satisfactory system responses can be achieved with the proposed DOBPH_∞C in Figure 3.6 compared with only the DOBC scheme in Figure 3.5. Figure 3.8 shows disturbance observer based on regional pole placement and D-stability theory can track the unknown disturbances very well in a short time compared with the pure DOBC in Figure 3.7.

3.6   CONCLUSION

It has been well known that both H_∞ control and DOBC are effective robust control schemes against external disturbances and modeling perturbations. H_∞ control generally achieves attenuation performance with respect to norm-bounded disturbances. DOBC has been used to reject the influence of the disturbance with some known information. In this chapter, these two schemes are combined to form a novel composite framework, which can be abbreviated as DOB plus H_∞ (DOBPH_∞) control. A class of nonlinear continuous systems is considered, where the disturbance has been divided into two parts: one is supposed to be a norm-bounded vector; the other is described by an exogenous system with perturbations. With such a control scheme, the two different types of disturbance can be attenuated and rejected, respectively. The proposed approach improves the classical DOBC method given in [85].

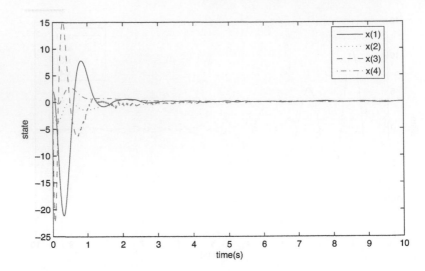

FIGURE 3.2   System performance using DOBPH_∞C with disturbances for known nonlinearity.

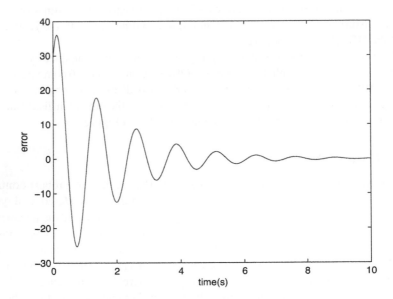

FIGURE 3.3   Estimation error for disturbances using DOBC for known nonlinearity.

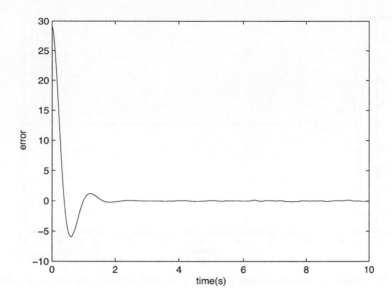

FIGURE 3.4   Estimation error for disturbances using DOBPH_∞C for known nonlinearity.

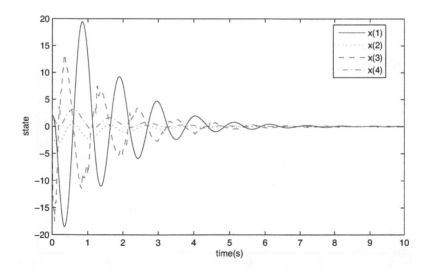

FIGURE 3.5   System performance using DOBC with disturbances for unknown nonlinearity.

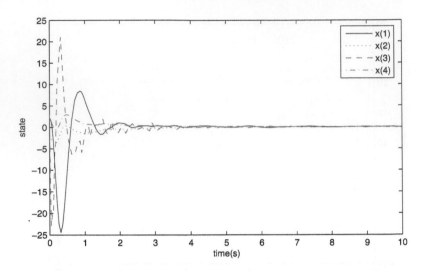

FIGURE 3.6   System performance using DOBPH_∞C with disturbances for unknown nonlinearity.

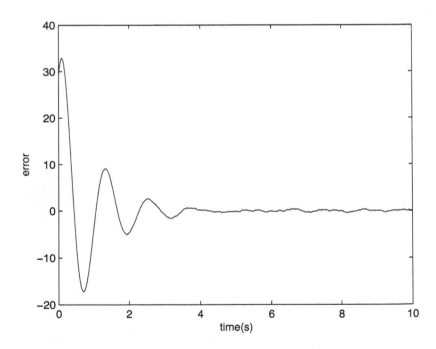

FIGURE 3.7   Estimation error for disturbances using DOBC for unknown nonlinearity.

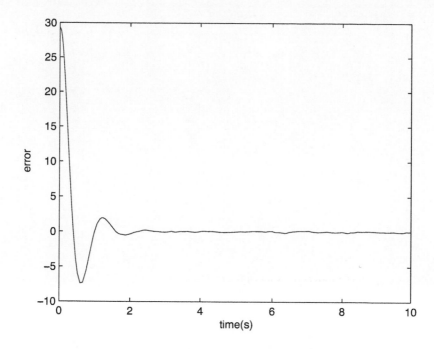

FIGURE 3.8   Estimation error for disturbances using DOBPH_∞C for unknown nonlinearity.

4 Composite DOBC and TSM Control for Nonlinear Systems with Disturbances

4.1 INTRODUCTION

Uncertainties cannot be included in such an exogenous model for disturbance, which restricts the application of the results in [31, 85, 144] (also in all other literature where the exogenous system is assumed to have a precise model). In addition, variable structure systems (VSSs) are well known for their robustness to system parameter variations and external disturbances [141, 196]. A particular interesting aspect of VSS is the sliding mode control, which is designed to drive and constrain the system states to lie within a neighborhood of the prescribed switching manifolds which exhibit desired dynamics. When in the sliding mode, the closed-loop response becomes totally insensitive to both internal parameter uncertainties and external disturbances. Compared with linear hyperplane based on sliding modes, terminal sliding mode (TSM) offers some superior properties such as fast, finite time convergence (e.g., [54, 192, 207]). The TSM controller is particularly useful for high precision control as it speeds up the rate of convergence near an equilibrium point.

It is well known that the control approach is mainly based on complete knowledge of a system's model, while in this chapter, the research is focused on control in the presence of disturbance. Composite disturbance is considered in MIMO continuous systems and includes two parts. One part is the bounded vector in H_2-norm context. The other part is supposed to be generated by an exogenous system, which is not required to be neutral stable as in the output regulation theory. Especially, the exogenous system is also supposed to have the modeling perturbation. A reduced-order observer based on regional pole placement and D-stability theory is structured for the estimation of the disturbance. A composite control scheme with DOBC and terminal sliding mode control is presented so that the disturbance with the exogenous system can be estimated and compensated for, and the disturbance with the bounded norm can be attenuated by terminal sliding mode control law. Two simulation examples for a flight control system and a hard disk drive actuator demonstrate the superiority of the proposed composite control scheme.

4.2 FORMULATION OF THE PROBLEM

The following continuous MIMO system with nonlinear dynamics and two types of disturbances is considered

$$
\begin{cases}
\dot{x}(t) &= G_0 x(t) + F_{01} f_{01}(x(t),t) \\
&\quad + H_0[u(t) + d_0(t)] + H_1 d_1(t) \\
y(t) &= C_0 x(t) + F_{02} f_{02}(x(t),t) \\
&\quad + D_0 d_0(t) + D_1 d_1(t)
\end{cases}
\tag{4.1}
$$

where $x(t) \in R^n$, $u(t) \in R^m (m < n)$, $y(t) \in R^l$ is state, control input and measurement output, respectively. $G_0 \in R^{n \times n}, H_1 \in R^{n \times l}, C_0 \in R^{l \times n}, D_0 \in R^{l \times m}, D_1 \in R^{l \times l}$ are the coefficient matrices. $H_0 \in R^{n \times m}$ satisfies $rank(H_0) = m$. F_{01} and F_{02} are the corresponding weighting matrices. $f_{01}(x(t),t)$ and $f_{02}(x(t),t)$ are nonlinear functions which are supposed to satisfy bounded conditions described as Assumption 4.2. $d_0(t) \in R^m$ is supposed to be described by an exogenous system in Assumption 4.1, which can represent the constant and the harmonic signals with modeling perturbations. $d_1(t) \in R^l$ is the external disturbance in the H_2-norm.

Assumption 4.1 *The disturbance $d_0(t)$ in the control input path can be formulated by the following exogenous system*

$$
\begin{cases}
\dot{w}(t) &= W w(t) + H_2 \delta(t) \\
d_0(t) &= V w(t)
\end{cases}
\tag{4.2}
$$

where W, H_2 and V are known matrices. $\delta(t)$ is the additional disturbance which results from the perturbations and uncertainties in the exogenous system.

The following assumption is required to describe the nonlinear dynamics in [192].

Assumption 4.2 *The nonlinear functions satisfy*

$$
\begin{cases}
\|f_{0i}(x(t),t)\| &\leq \beta_{0i} + \beta_{1i} \|x((t),t)\| \\
f_{0i}(0,t) &= 0, \quad i = 1,2
\end{cases}
\tag{4.3}
$$

where β_{0i}, β_{1i} are positive numbers.

The following assumption is a necessary condition for the DOBC formulation.

Assumption 4.3 *(G_0, H_0) is controllable; $(W, H_0 V)$ is observable.*

Suppose that the system state is available. If $d_0(t)$ is considered as a part of the augmented state, then a reduced-order observer is needed. In this chapter, we construct the following reduced-order disturbance observer for the case with known nonlinearity and unknown nonlinearity, respectively. And then we design a composite controller with the disturbance observer and a TSM controller so that the disturbances can be rejected and attenuated simultaneously, while the desired dynamic performances can be guaranteed for MIMO continuous systems in finite time.

4.3 COMPOSITE DOBC AND TSM CONTROL FOR THE CASE WITH KNOWN NONLINEARITY

In this section, we suppose that $f_{01}(x(t),t)$ and $f_{02}(x(t),t)$ are given and Assumptions 4.1-4.3 hold. When all states of the system are available, it is unnecessary to estimate the states, and only the estimation of the disturbance needs to be included.

4.3.1 DISTURBANCE OBSERVER

In this section, the disturbance observer is formulated as

$$
\begin{cases}
\hat{d}_0(t) &= V\hat{w}(t) \\
\hat{w}(t) &= v(t) - Lx(t) \\
\dot{v}(t) &= (W + LH_0V)(v(t) - Lx(t)) + L(G_0x(t) \\
& \quad + F_{01}f_{01}(x(t),t) + H_0u(t))
\end{cases}
\tag{4.4}
$$

where $\hat{w}(t)$ is the estimation of $w(t)$, and $v(t)$ is the auxiliary vector as the state of the observer. The estimation error is denoted as $e_w(t) = w(t) - \hat{w}(t)$. Based on (4.1), (4.2) and (4.4), it is shown that the error dynamics satisfies

$$
\dot{e}_w(t) = (W + LH_0V)e_w(t) + H_2\delta(t) + LH_1d_1(t)
\tag{4.5}
$$

and is further transformed to give

$$
\begin{cases}
\dot{e}_w(t) &= (W + LH_0V)e_w(t) + Hd(t) \\
z(t) &= e_w(t)
\end{cases}
\tag{4.6}
$$

where $z(t)$ is denoted as the reference output, and

$$
H = \begin{bmatrix} H_2 & LH_1 \end{bmatrix}, \quad d(t) = \begin{bmatrix} \delta^T(t) & d_1^T(t) \end{bmatrix}^T
\tag{4.7}
$$

Based on regional pole placement and H_∞ theory, the following result can be obtained.

Theorem 4.1

For parameters $\theta_i > 0$ ($i = 1, 2, ..., n$), suppose that there exists the Lyapunov matrix $P > 0$ and matrix T, satisfying

$$
\min \gamma > 0
\tag{4.8}
$$

$$\begin{bmatrix} PW + W^T P + TH_0 V + V^T H_0^T T^T + I & PH_2 & TH_1 \\ * & -\gamma^2 I & 0 \\ * & * & -\gamma^2 I \end{bmatrix} < 0 \qquad (4.9)$$

$$PW + W^T P + TH_0 V + V^T H_0^T T^T + P\Theta < 0 \qquad (4.10)$$

then by selecting $L = P^{-1}T, \Theta = diag(\theta_i)$, the error system (4.6) is asymptotically stable in the absence of $d(t)$ and satisfies $\|z(t)\|_2 \le \gamma \|d(t)\|_2$, where γ represents the level of disturbance attenuation. ∎

Proof. Consider the following Lyapunov function

$$V(t) = e_\omega^T(t) P e_\omega(t)$$

In order to testify that system (4.6) is asymptotically stable with a disturbance attenuation $\gamma > 0$, the associated function can be considered as

$$J(t) = \int_0^t [z^T(t)z(t) - \gamma^2 d^T(t)d(t) + \dot{V}(t)]dt$$

Defining

$$H(t) = z^T(t)z(t) - \gamma^2 d^T(t)d(t) + \dot{V}(t)$$

then it is shown that

$$\begin{aligned} H(t) &= z^T(t)z(t) - \gamma^2 d^T(t)d(t) + e_\omega^T(t)P\dot{e}_\omega(t) + \dot{e}_\omega^T(t)Pe_\omega^T(t) \\ &= z^T(t)z(t) - \gamma^2 d^T(t)d(t) + e_\omega^T(t)P[(W + LH_0 V)e_w(t) + Hd(t)] \\ &\quad + [(W + LH_0 V)e_w(t) + Hd(t)]^T Pe_\omega(t) \\ &= [e_\omega^T(t) \quad d^T(t)] \begin{bmatrix} \Theta_1 & PH \\ H^T P & -\gamma^2 I \end{bmatrix} \begin{bmatrix} e_\omega(t) \\ d(t) \end{bmatrix} \end{aligned} \qquad (4.11)$$

where

$$\Theta_1 = P(W + LH_0 V) + (W + LH_0 V)^T P + I$$

Substituting (4.7) into (4.11) yields

$$H(t) = [e_\omega^T(t) \quad \delta^T(t) \quad d_1^T(t)] \begin{bmatrix} \Theta_1 & PH_2 & PLH_1 \\ * & -\gamma^2 I & 0 \\ * & * & -\gamma^2 I \end{bmatrix} \begin{bmatrix} e_\omega(t) \\ \delta(t) \\ d_1(t) \end{bmatrix} \qquad (4.12)$$

Under the zero initial condition, it can be shown that $J(t) < 0$ and $\|z(t)\|_2 \le \gamma \|d(t)\|_2$, if $H(t) < 0$ holds. Based on (4.12), we can get (4.8), (4.9) after letting $T =$

PL. According to regional pole placement and D-stability theory in [38], choose regional pole $\Theta > 0$ for systems (4.6), then if $X > 0$, it can be verified that

$$(W + LH_0V)X + X(W + LH_0V)^T + \Theta X < 0 \tag{4.13}$$

Then the pole will be assigned in following LMI region D.

$$D = \{s \in C : \Theta + sM + \bar{s}M^T < 0\}$$

(4.10) is obtained by defining $X = P^{-1}$, the proof is completed.   □
In this section, we construct reduced-order disturbance observer (4.4) to estimate disturbance $d_0(t)$. In the following section, we will design a composite controller with the disturbance observer and a TSM controller so that the disturbance can be rejected and attenuated, and simultaneously the desired dynamic performances can be guaranteed for system (4.1).

4.3.2 COMPOSITE DOBC AND TSM CONTROL

The intention in this section is to design a control strategy to guarantee that system (4.1) is stable in Lyapunov's sense, that is, to regulate the system states from any initial states $x(0) \neq 0$ to a neighborhood of the equilibrium points.

The structure of the controller is formulated as

$$u(t) = -\hat{d}_0(t) + u_{tsm}(t) \tag{4.14}$$

where $u_{tsm}(t)$ is TSM controller. Substituting (4.14) into (4.1), it can be shown that

$$\dot{x}(t) = G_0x(t) + H_0Ve_w(t) + H_0u_{tsm}(t) + F_{01}f_{01}(x(t),t) + H_1d_1(t) \tag{4.15}$$

Considering $rank(H_0) = m$, there exists a nonsingular linear transformation $\zeta(t) = Mx(t)$, such that system (4.15) can be transformed as

$$\dot{\zeta}(t) = \bar{G}_0\zeta(t) + \bar{H}_0u_{tsm}(t) + MF_{01}f_{01}(x(t),t) + \bar{f} \tag{4.16}$$

where

$$\zeta(t) = \begin{bmatrix} \zeta_1(t) \\ \zeta_2(t) \end{bmatrix}, \quad MF_{01} = \begin{bmatrix} M_{01} \\ M_{02} \end{bmatrix},$$

$$\bar{G}_0 = MG_0M^{-1} = \begin{bmatrix} \bar{G}_{01} & \bar{G}_{02} \\ \bar{G}_{03} & \bar{G}_{04} \end{bmatrix},$$

$$\bar{H}_0 = MH_0 = \begin{bmatrix} 0 \\ \bar{H}_{01} \end{bmatrix}, \quad \bar{H}_1 = MH_1,$$

$$\bar{f} = \bar{H}_0Ve_w(t) + \bar{H}_1d_1(t) = \begin{bmatrix} \bar{f}_1 \\ \bar{f}_2 \end{bmatrix},$$

$$\|\bar{f}_1\| \le k_1, \quad \|\bar{f}_2\| \le k_2 \tag{4.17}$$

and $\zeta_1(t) \in R^{n-m}$, $\zeta_2(t) \in R^m$, $\bar{G}_{01} \in R^{(n-m)\times(n-m)}$, $\bar{G}_{02} \in R^{(n-m)\times m}$, $\bar{G}_{03} \in R^{m\times m}$, $\bar{G}_{04} \in R^{m\times(n-m)}$, $\bar{f}_1 \in R^{n-m}$, $\bar{f}_2 \in R^m$. k_1, k_2 are known constants, $\bar{H}_{01} \in R^{m\times m}$ is nonsingular.

Then, Equation (4.16) can be formulated as

$$\begin{cases} \dot{\zeta}_1(t) = \bar{G}_{01}\zeta_1(t) + \bar{G}_{02}\zeta_2(t) + M_{01}f_{01}(x(t),t) + \bar{f}_1 \\ \dot{\zeta}_2(t) = \bar{G}_{03}\zeta_1(t) + \bar{G}_{04}\zeta_2(t) + \bar{H}_{01}u_{tsm}(t) \\ \qquad\qquad + M_{02}f_{01}(x(t),t) + \bar{f}_2 \end{cases} \tag{4.18}$$

For (4.18), according to [141], the TSM is defined as

$$S(t) = C_1\zeta_1(t) + C_2\zeta_2(t) + C_3\zeta_1^{\frac{q}{p}}(t) \tag{4.19}$$

where $C_1 \in R^{m\times(n-m)}$, $C_2 \in R^{m\times m}$, $C_3 \in R^{m\times(n-m)}$ are design parameter matrices, C_2 is nonsingular, p and q are odd integers and satisfy

$$q < p < 2q \tag{4.20}$$

The vector $\zeta_1^{\frac{q}{p}}(t)$ is defined as follows:

$$\zeta_1^{\frac{q}{p}}(t) = \begin{bmatrix} \zeta_{11}^{\frac{q}{p}}, & \zeta_{12}^{\frac{q}{p}}, & \dots, & \zeta_{1n-m}^{\frac{q}{p}} \end{bmatrix}^T$$

For MIMO continuous systems (4.18), the TSM controller based on disturbance observer is designed as

$$u(t) = u_0 + u_{tsm}; \quad u_0 = -\hat{d}_0(t); \quad u_{tsm} = u_1 + u_2; \tag{4.21}$$

$$u_1 = \begin{cases} -(C_2\bar{H}_{01})^{-1}\{(C_1\bar{G}_{01} + C_2\bar{G}_{03})\zeta_1(t) + (C_1\bar{G}_{02} + C_2\bar{G}_{04})\zeta_2(t) \\ \quad + \frac{q}{p}C_3 diag(\zeta_1^{\frac{q}{p}-1})[\bar{G}_{01}\zeta_1(t) + \bar{G}_{02}\zeta_2(t) + M_{01}f_{01}(x(t),t)] \\ \quad + C_1 M_{01}f_{01}(x(t),t) + C_2 M_{02}f_{01}(x(t),t) + \Phi S^r\} \\ \qquad\qquad for\ S \neq 0\ and\ \zeta_1(t) \neq 0 \\ -(C_2\bar{H}_{01})^{-1}[C_2\bar{G}_{04}\zeta_2(t) + C_2 M_{02}f_{01}(x(t),t) + \Phi S^r] \\ \qquad\qquad for\ S \neq 0\ and\ \zeta_1(t) = 0 \end{cases} \tag{4.22}$$

$$u_2 = \begin{cases} -(C_2\bar{H}_{01})^{-1}\rho_1 \frac{S}{\|S\|} & for\ S \neq 0\ and\ \zeta_1(t) \neq 0 \\ -(C_2\bar{H}_{01})^{-1}\rho_2 \frac{S}{\|S\|} & for\ S \neq 0\ and\ \zeta_1(t) = 0 \end{cases} \tag{4.23}$$

where $\Phi = diag(\phi_1, \cdots, \phi_m)$, $\phi_i > 0, i = 1, \cdots, m$, are designed parameters; r is a constant, which satisfies $r = \frac{g_1}{g_2}$, where g_1 and g_2 are two odd integers, and $g_1 < g_2$, so it is obtained that $0 < r < 1$. ρ_1, ρ_2 are positive scalars, which are designed in Theorem 4.2.

Remark 4.1 In view of a nonsingular linear transformation $\zeta(t) = Mx(t)$, the TSM controller designed by (4.21)-(4.23) for system (4.18) is the same as that for system (4.1).

In order to prove the following theorem, a lemma in [192] should be considered, which is described as follows.

Lemma 4.1

Assume that a continuous, positive-definite function $V(t)$ satisfies the following differential inequality

$$\dot{V}(t) \leq -\alpha V^{\eta}(t), \quad \forall t \geq t_0, \quad V(t) \geq 0$$

where $\alpha > 0$, $0 < \eta < 1$ are constants. Then, for any given t_0, $V(t)$ satisfies the following inequality

$$V^{1-\eta}(t) \leq V^{1-\eta}(t_0) - \alpha(1-\eta)(t-t_0), \quad t_0 \leq t \leq t_r$$

and satisfies the following equality

$$V(t) = 0, \quad t \geq t_r$$

with t_r given by

$$t_r = t_0 + \frac{V^{1-\eta}(t_0)}{\alpha(1-\eta)}$$

∎

The following theorem will be proposed to design TSM controller (4.21)-(4.23) to guarantee that MIMO continuous systems (4.1) will converge into a limited region around the equilibrium point.

Theorem 4.2

For MIMO continuous system (4.1) with TSM manifold (4.19), if the TSM controller is designed as (4.21)-(4.23) and the terminal sliding mode parameter matrices satisfy

$$\bar{G}_{01} - \bar{G}_{02}C_2^{-1}C_1 = -Q_1 \tag{4.24}$$

$$\bar{G}_{02}C_2^{-1}C_3 = \Lambda \tag{4.25}$$

then the states of system (4.1) will reach the terminal sliding mode $S(t) = 0$ in finite time t_{r1}, and converge to the neighborhood of the equilibrium point Ω^* in finite time t_{r2}. Where Q_1 is a positive-definite matrix, $\Lambda = diag[\Lambda_1, \Lambda_2, ... \Lambda_{n-m}]$, and $\Lambda_i > 0$. ρ_1 and ρ_2 are positive scalars and satisfy the following inequality:

$$\rho_1 \geq k_1 (\|C_1\| + \frac{q}{p}\|C_3\|\|diag(\zeta_1^{(q/p-1)}(t))\|) + k_2\|C_2\|,$$

$$\rho_2 \geq k_2\|C_2\| \tag{4.26}$$

and t_{r1}, t_{r2}, Ω^* are given by

$$t_{r1} = t_0 + \frac{V_1^{1-\eta_1}(t_0)}{\alpha_1(1-\eta_1)}, \quad t_{r2} = t_{r1} + \frac{V_2^{1-\eta_2}(t_0)}{\alpha_2(1-\eta_2)} \tag{4.27}$$

$$\Omega^* = \left\{ x(t): \|x(t)\| \leq \|M^{-1}\| \sqrt{(\frac{\bar{k}_1}{\lambda_{min}(Q_1)})^2 + (\|C_2^{-1}C_3\|\|\zeta_1\| + \|C_3\|\|\zeta_1^{\frac{q}{p}}\|)^2} \right\},$$

$$\bar{k}_1 = \|M_{01}f_{01}(x(t),t)\| + k_1 \tag{4.28}$$

■

Proof. Differentiating $S(t)$ in (4.19) with respect to time along the system (4.18), and substituting the control laws (4.21)-(4.23) into it, it can be verified that
(a) for $S \neq 0$ and $\zeta_1(t) \neq 0$,

$$\begin{aligned}
\dot{S}(t) &= C_1\dot{\zeta}_1(t) + C_2\dot{\zeta}_2(t) + \frac{q}{p}C_3 diag(\zeta_1^{(q/p-1)}(t))\dot{\zeta}_1(t) \\
&= (C_1 + \frac{q}{p}C_3 diag(\zeta_1^{(q/p-1)}(t)))[\bar{G}_{01}\zeta_1(t) + \bar{G}_{02}\zeta_2(t) \\
&\quad + M_{01}f_{01}(x(t),t) + \bar{f}_1] + C_2[\bar{G}_{03}\zeta_1(t) \\
&\quad + \bar{G}_{04}\zeta_2(t) + \bar{H}_{01}u_{tsm}(t) + M_{02}f_{01}(x(t),t) + \bar{f}_2] \\
&= (C_1\bar{G}_{01} + C_2\bar{G}_{03})\zeta_1(t) + (C_1\bar{G}_{02} + C_2\bar{G}_{04})\zeta_2(t) \\
&\quad + \frac{q}{p}C_3 diag(\zeta_1^{\frac{q}{p}-1})[\bar{G}_{01}\zeta_1(t) + \bar{G}_{02}\zeta_2(t) + M_{01}f_{01}(x(t),t)] \\
&\quad + C_1 M_{01}f_{01}(x(t),t) + C_2 M_{02}f_{01}(x(t),t) + C_2\bar{H}_{01}u_{tsm}(t) \\
&\quad + [C_1 + \frac{q}{p}C_3 diag(\zeta_1^{(q/p-1)}(t))]\bar{f}_1 + C_2\bar{f}_2 \\
&= -\Phi S^r - \rho_1\frac{S}{\|S\|} + [C_1 + \frac{q}{p}C_3 diag(\zeta_1^{(q/p-1)}(t))]\bar{f}_1 + C_2\bar{f}_2 \tag{4.29}
\end{aligned}$$

Consider a candidate Lyapunov function $V_1(t)$ as follows

$$V_1(t) = \frac{1}{2}S^T S$$

By differentiating $V_1(t)$ with respect to time and substituting (4.29) into it, it can be shown that

$$
\begin{aligned}
\dot{V}_1(t) &= S^T \dot{S} \\
&= -S\Phi S^r + S^T [C_1 + \frac{q}{p}C_3 diag(\zeta_1^{(q/p-1)}(t))]\bar{f}_1 \\
&\quad -\rho_1 \|S\| + S^T C_2 \bar{f}_2
\end{aligned}
$$

Considering

$$S\Phi S^r = \sum_{i=1}\Phi_i S_i^{r+1} \geq \min_{i=1}(\Phi_i)[(\sum_{i=1}S_i^2)^{\frac{r+1}{2}}] = \min_{i=1}(\Phi_i)\|S\|^{r+1}$$

It can be verified that

$$
\begin{aligned}
\dot{V}_1(t) \leq\ & -\min_{i=1}(\Phi_i)\|S\|^{r+1} - \|S\|\{\rho_1 - k_1(\|C_1\| \\
& -\frac{q}{p}\|C_3\|\|diag(\zeta_1^{(q/p-1)}(t))\|) - k_2\|C_2\|\}
\end{aligned}
$$

Based on (4.26), it can be obtained

$$
\begin{aligned}
\dot{V}_1(t) &\leq -\min_{i=1}(\Phi_i)\|S\|^{r+1} = -\min_{i=1}(\Phi_i)(2V(t))^{\frac{r+1}{2}} \\
&= -\alpha_1 V_1^{\eta_1}(t)
\end{aligned}
$$

where

$$\alpha_1 = 2^{\frac{r+1}{2}}\min_{i=1}(\Phi_i), \quad \eta_1 = \frac{r+1}{2} < 1.$$

According to Lemma 4.1, systems (4.18) will reach the TSM manifold $S = 0$ in finite time t_{r1} which is given by

$$t_{r1} = t_0 + \frac{V_1^{1-\eta_1}(t_0)}{\alpha_1(1-\eta_1)} \tag{4.30}$$

(b) for $S \neq 0$ and $\zeta_1(t) = 0$. The terminal sliding mode can be expressed as

$$S(t) = C_2\zeta_2(t) \tag{4.31}$$

Similar to the proof of (a), based on Lemma 4.1 and (4.21)-(4.23), it is shown that systems (4.18) will reach the TSM manifold $S = 0$ in finite time t_{r1} given by (4.30).

Based on the above proof and in view of a nonsingular linear transformation $\zeta(t) = Mx(t)$, systems (4.1) will also reach the TSM manifold $S = 0$ in finite time t_{r1} simultaneously.

When confined to the manifold $S = 0$, the dynamics of the system (4.1) and (4.18) will be determined by the sliding mode manifold. According to terminal sliding mode (4.19), the state vector $\zeta_2(t)$ can be expressed as

$$\zeta_2(t) = -C_2^{-1}(C_1\zeta_1(t) + C_3\zeta_1^{\frac{q}{p}}(t)) \tag{4.32}$$

Substituting (4.32) with (4.18), it can be shown that

$$\begin{aligned}
\dot{\zeta}_1(t) &= (\bar{G}_{01} - \bar{G}_{02}C_2^{-1}C_1)\zeta_1(t) - \bar{G}_{02}C_2^{-1}C_3\zeta_1^{\frac{q}{p}}(t) \\
&\quad + M_{01}f_{01}(x(t),t) + \bar{f}_1
\end{aligned} \tag{4.33}$$

For system (4.33), consider a Lyapunov function candidate

$$V_2(t) = \frac{1}{2}\zeta_1(t)^T\zeta_1(t)$$

Based on (4.24) and (4.25), along with the trajectories of (4.33), it can be verified that

$$\begin{aligned}
\dot{V}_2(t) &= \zeta_1(t)^T\dot{\zeta}_1(t) = \zeta_1(t)^T[-Q_1\zeta_1(t) - \Lambda\zeta_1^{\frac{q}{p}}(t) \\
&\quad + M_{01}f_{01}(x(t),t) + \bar{f}_1] \\
&= -\zeta_1(t)^TQ_1\zeta_1(t) - \zeta_1(t)^T\Lambda\zeta_1^{\frac{q}{p}}(t) \\
&\quad + \zeta_1(t)^T(M_{01}f_{01}(x(t),t) + \bar{f}_1)
\end{aligned} \tag{4.34}$$

Considering

$$\begin{aligned}
\zeta_1(t)^T\Lambda\zeta_1^{\frac{q}{p}}(t) &= \sum_{i=1}^{n-m}\Lambda_i\zeta_{1i}^{\frac{q}{p}+1}(t) \geq \min_i(\Lambda_i)[(\sum_{i=1}^{n-m}\zeta_{1i}^2)^{\frac{q}{p}+1}{2}] \\
&= \min_i(\Lambda_i)\|\zeta_1(t)\|^{\frac{q}{p}+1}
\end{aligned} \tag{4.35}$$

Based on (4.28), it can be verified that

$$\begin{aligned}
\dot{V}_2(t) &\leq -\lambda_{min}(Q_1)\|\zeta_1(t)\|^2 - \min_i(\Lambda_i)\|\zeta_1(t)\|^{\frac{q}{p}+1} \\
&\quad + \|\zeta_1(t)\|(\|M_{01}f_{01}(x(t),t)\| + k_1) \\
&= -\min_i(\Lambda_i)\|\zeta_1(t)\|^{\frac{q}{p}+1} - \|\zeta_1(t)\|(\lambda_{min}(Q_1)\|\zeta_1(t)\| \\
&\quad - \|M_{01}f_{01}(x(t),t)\| - k_1) \\
&= -\alpha_2 V_2^{\eta_2} - \|\zeta_1(t)\|(\lambda_{min}(Q_1)\|\zeta_1(t)\| - \bar{k}_1)
\end{aligned} \tag{4.36}$$

where

$$\alpha_2 = 2^{\frac{q}{p}+1}{2} \min_i(\Lambda_i), \quad \eta_2 = \frac{\frac{q}{p}+1}{2} < 1$$

When system state satisfies $\|\zeta_1(t)\| > \frac{\bar{k}_1}{\lambda_{min}(Q_1)}$, it can be obtained

$$\dot{V}_2(t) \leq -\alpha_2 V_2^{\eta_2}$$

According to Lemma 4.1, the states of system $\zeta_1(t)$ will converge to the neighborhood of the equilibrium point Ω_1 in finite time t_{r2}. Where

$$\Omega_1 = \left\{ \zeta_1 : \|\zeta_1\| \leq \frac{\bar{k}_1}{\lambda_{min}(Q_1)} \right\}, \quad t_{r2} = t_{r1} + \frac{V_2^{1-\eta_2}(t_0)}{\alpha_2(1-\eta_2)} \tag{4.37}$$

According to (4.32), when the states ζ_1 of system (4.18) converge into Ω_1, the state ζ_2 will also converge into a region Ω_2 which is determined by ζ_1 as follow

$$\Omega_2 = \left\{ \zeta_2 : \|\zeta_2\| \leq \|C_2^{-1}C_3\|\|\zeta_1\| + \|C_3\|\|\zeta_1^{\frac{q}{p}}\| \right\}$$

Based on nonsingular linear transformation $\zeta(t) = Mx(t)$, it can be verified that

$$\begin{aligned}\|x(t)\| &= \|M^{-1}\|\|\zeta(t)\| = \|M^{-1}\|\sqrt{\|\zeta_1(t)\|^2 + \|\zeta_2(t)\|^2} \\ &\leq \|M^{-1}\|\sqrt{(\frac{\bar{k}_1}{\lambda_{min}(Q_1)})^2 + (\|C_2^{-1}C_3\|\|\zeta_1\| + \|C_3\|\|\zeta_1^{\frac{q}{p}}\|)^2}\end{aligned}$$

Namely, the states of system (4.1) will converge to the neighborhood of the equilibrium point $\Omega^\star$ in finite time t_{r2}. Where

$$\Omega^\star = \left\{ x(t) : \|x(t)\| \leq \|M^{-1}\|\sqrt{(\frac{\bar{k}_1}{\lambda_{min}(Q_1)})^2 + (\|C_2^{-1}C_3\|\|\zeta_1\| + \|C_3\|\|\zeta_1^{\frac{q}{p}}\|)^2} \right\}$$

The proof is completed. □

With the above formulation, it can be seen that the composite system is combined with (4.5) and (4.15). One is the error dynamical system for the estimation of the disturbance governed by the exogenous system. The other results from the original system by using the disturbance rejection term in control input and merging two types of disturbance into an augmented one. The main objective is to design an observer to estimate the first type of disturbance, and construct a composite TSM controller so that the states of system (4.1) will also reach the TSM manifold $S = 0$ in finite time. The new control scheme combines the classical DOBC with TSM control, which can be called the "DOBC plus TSM control," and abbreviated as DOBPTSMC.

4.4 COMPOSITE DOBC AND TSM CONTROL FOR THE CASE WITH UNKNOWN NONLINEARITY

In this section, we suppose Assumptions 4.1-4.3 hold, but nonlinear functions $f_{01}(x(t),t)$ and $f_{02}(x(t),t)$ are unknown. Different from Section 4.3, $f_{01}(x(t),t)$ is unavailable in observer design.

4.4.1 DISTURBANCE OBSERVER

In this section, the disturbance observer is formulated as

$$\begin{cases} \dot{\hat{d}}_0(t) & = & V\hat{w}(t) \\ \hat{w}(t) & = & v(t) - Lx(t) \\ \dot{v}(t) & = & (W + LH_0V)(v(t) - Lx(t)) + L(G_0x(t) + H_0u(t)) \end{cases} \tag{4.38}$$

Compared with (4.5), the estimation error $e_w(t) = w(t) - \hat{w}(t)$ satisfies

$$\dot{e}_w(t) = (W + LH_0V)e_w(t) + LF_{01}f_{01} + H_2\delta(t) + LH_1d_1(t) \tag{4.39}$$

and is further transformed to give

$$\begin{cases} \dot{e}_w(t) & = & (W + LH_0V)e_w(t) + Hd(t) \\ z(t) & = & e_\omega(t) \end{cases} \tag{4.40}$$

where $z(t)$ is the reference output, and

$$H = \begin{bmatrix} H_2 & LH_1 & LF_{01} \end{bmatrix}, \quad d(t) = \begin{bmatrix} \delta^T(t) & d_1^T(t) & f_{01}^T \end{bmatrix}^T \tag{4.41}$$

Similar to Theorem 4.1, based on regional pole placement and H_∞ theory, the following result can be obtained.

Theorem 4.3

For parameters $\theta_i > 0, i = 1, 2, ..., n$, suppose that there exist the Lyapunov matrix $P > 0$ and matrix T, satisfying

$$\min \gamma > 0 \tag{4.42}$$

$$\begin{bmatrix} PW + W^T P + TH_0V + V^T H_0^T T^T + I & PH_2 & TH_1 & TF_{01} \\ * & -\gamma^2 I & 0 & 0 \\ * & * & -\gamma^2 I & 0 \\ * & * & * & -\gamma^2 I \end{bmatrix} < 0 \tag{4.43}$$

$$PW + W^T P + TH_0V + V^T H_0^T T^T + P\Theta < 0 \tag{4.44}$$

then by selecting $L = P^{-1}T$, $\Theta = diag(\theta_i)$, the error system (4.40) is asymptotically stable in the absence of $d(t)$ and satisfies $\|z(t)\|_2 \le \gamma \|d(t)\|_2$, where γ represents the level of disturbance attenuation. ∎

Proof. The proof procedure can be given similarly to that of the proof for Theorem 4.1. □

4.4.2 COMPOSITE DOBC AND TSM CONTROL

For MIMO continuous systems (4.1), the TSM controller based on disturbance observer is designed as

$$u(t) = u_0 + u_v; \quad u_{tsm} = u_1 + u_2; \quad u_0 = -\hat{d}_0(t); \tag{4.45}$$

$$u_1 = \begin{cases} -(C_2\bar{H}_{01})^{-1}\{(C_1\bar{G}_{01} + C_2\bar{G}_{03})\zeta_1(t) + (C_1\bar{G}_{02} + C_2\bar{G}_{04})\zeta_2(t) \\ \quad + \frac{q}{p}C_3 diag(\zeta_1^{\frac{q}{p}-1})[\bar{G}_{01}\zeta_1(t) + \bar{G}_{02}\zeta_2(t)] + \Phi S^r\} \\ \qquad for\ S \ne 0\ and\ \zeta_1(t) \ne 0 \\ -(C_2\bar{H}_{01})^{-1}[C_2\bar{G}_{04}\zeta_2(t) + \Phi S^r] \\ \qquad for\ S \ne 0\ and\ \zeta_1(t) = 0 \end{cases} \tag{4.46}$$

$$u_2 = \begin{cases} -(C_2\bar{H}_{01})^{-1}\rho_1\frac{S}{\|S\|} & for\ S \ne 0\ and\ \zeta_1(t) \ne 0 \\ -(C_2\bar{H}_{01})^{-1}\rho_2\frac{S}{\|S\|} & for\ S \ne 0\ and\ \zeta_1(t) = 0 \end{cases} \tag{4.47}$$

where $\Phi = diag(\phi_1, \cdots, \phi_m)$, $\phi_i > 0$, $i = 1, \cdots, m$, are design parameters; r is a constant, which satisfies $r = \frac{g_1}{g_2}$, where g_1 and g_2 are two odd integers, and $g_1 < g_2$, so it is obtained that $0 < r < 1$. ρ_1, ρ_2 are positive scalars, which are designed in Theorem 4.2.

Similar to Theorem 4.2, the following result can be obtained.

Theorem 4.4

For MIMO continuous system (4.1) with TSM manifold (4.19), if the TSM control is designed as (4.45)-(4.47), and the terminal sliding mode parameter matrices satisfy

$$\bar{G}_{01} - \bar{G}_{02}C_2^{-1}C_1 = -Q_2 \tag{4.48}$$

$$\bar{G}_{02}C_2^{-1}C_3 = \Lambda \tag{4.49}$$

then the states of system (4.1) will reach the terminal sliding mode $S(t) = 0$ in finite time t_{r3}, and converge to the neighborhood of the equilibrium point Ω_1^* in finite time

t_{r4}. Where Q_2 is a positive-definite matrix, $\Lambda = diag[\Lambda_1, \Lambda_2, ... \Lambda_{n-m}]$, and $\Lambda_i > 0$. ρ_1 and ρ_2 are positive scalars and satisfy the following inequality:

$$\rho_1 \geq k_1'[\|C_1\| + \frac{q}{p}\|C_3\| \| diag(\zeta_1^{(q/p-1)}(t))\|] + k_2'\|C_2\|,$$

$$\rho_2 \geq k_2'\|C_2\| \tag{4.50}$$

where

$$k_1' = k_1 + \|M_{01}\|(\beta_{01} + \beta_{11}\|M^{-1}\zeta(t)\|)$$

$$k_2' = k_2 + \|M_{02}\|(\beta_{01} + \beta_{11}\|M^{-1}\zeta(t)\|)$$

and t_{r3}, t_{r4}, $\Omega_1^\star$ are given by

$$t_{r3} = t_0 + \frac{V_3^{1-\eta_3}(t_0)}{\alpha_3(1-\eta_3)}, \quad t_{r4} = t_{r3} + \frac{V_4^{1-\eta_4}(t_0)}{\alpha_4(1-\eta_4)} \tag{4.51}$$

$$\Omega_1^\star = \left\{ x(t): \ \|x(t)\| < \|M^{-1}\| \sqrt{(\frac{k_1'}{\lambda_{min}(Q_2)})^2 + (\|C_2^{-1}C_3\| \|\zeta_1\| + \|C_3\| \| \zeta_1^{\frac{q}{p}}\|)^2} \right\}$$

                                                       ∎

Proof. The proof procedure can be given similarly to that of the proof for Theorem 4.2.   □

Remark 4.2 The negative fractional powers existing in the TSM control (4.22) and (4.46) may cause the singularity problem around the equilibrium. Therefore, the two odd integers p and q in the terminal sliding mode (4.19) must be carefully selected according to (4.20). Moreover, some novel control strategy such as ([54, 207]) can be applied for further improvements.

Remark 4.3 If the control signal u_2 in (4.23) and (4.47) are replaced by the following control:

$$u_2^\varepsilon = \begin{cases} -(C_2\bar{H}_{01})^{-1}\rho_1 \frac{S}{\|S\|+\varepsilon} & for S \neq 0 \ and \ \zeta_1(t) \neq 0 \\ -(C_2\bar{H}_{01})^{-1}\rho_2 \frac{S}{\|S\|+\varepsilon} & for S \neq 0 \ and \ \zeta_1(t) = 0 \end{cases}$$

where ε is a small positive constant. Then, the control chattering has been reduced further.

4.5 SIMULATION EXAMPLES

4.5.1 SIMULATION ON JET TRANSPORT

A jet transport during cruise flight at Mach 0.8 and a 40,000-foot altitude can be described as

$$\dot{x}(t) = G_0 x(t) + F_{01} f_{01}(x(t), t) + H_0 [u(t) + d_0(t)] + H_1 d_1(t)$$

$$y(t) = C_0 x(t) + F_{02} f_{02}(x(t), t) + D_0 d_0(t) + D_1 d_1(t)$$

where $x(t) = \begin{bmatrix} x_1(t) & x_2(t) & x_3(t) & x_4(t) \end{bmatrix}^T$ and $x_1(t)$ is the sideslip angle, $x_2(t)$ is the yaw rate, $x_3(t)$ is the roll rate, $x_4(t)$ is the bank angle, $u(t)$ is the rudder and aileron deflections. The coefficient matrices of aircraft model in control system toolbox of MATLAB® 7.0 are given by

$$G_0 = \begin{bmatrix} -0.0558 & -0.9968 & 0.0802 & 0.0415 \\ 0.5980 & -0.1150 & -0.0318 & 0 \\ -3.0500 & 0.3880 & -0.4650 & 0 \\ 0 & 0.0805 & 1.0000 & 0 \end{bmatrix};$$

$$F_{01} = \begin{bmatrix} 0 \\ 0 \\ 5 \\ 0 \end{bmatrix}; \quad H_0 = \begin{bmatrix} 0.0073 & 0 \\ -0.4750 & 0.0077 \\ 0.1530 & 0.1430 \\ 0 & 0 \end{bmatrix}; \quad H_1 = \begin{bmatrix} 0.1 \\ 0 \\ -3 \\ 0.1 \end{bmatrix};$$

$$C_0 = \begin{bmatrix} 0 & 1 & 0 & 0 \\ 0 & 0 & 0 & 1 \end{bmatrix}; \quad F_{02} = \begin{bmatrix} 0.001 \\ 0.001 \end{bmatrix};$$

$$D_0 = \begin{bmatrix} 0.001 & 0 \\ 0 & 0.001 \end{bmatrix}; \quad D_1 = \begin{bmatrix} 0.001 \\ 0.001 \end{bmatrix};$$

$d_0(t)$ is assumed to be an unknown disturbance described by (4.2) with

$$W = \begin{bmatrix} 0 & 5 \\ -5 & 0 \end{bmatrix}; \quad V = \begin{bmatrix} 25 & 0 \\ 0 & 25 \end{bmatrix}; \quad H_2 = \begin{bmatrix} 0.1 \\ 0.1 \end{bmatrix}$$

where 5 in W represents the frequency of the signal. [85] pointed out that if the frequency is perturbed, the pure DOBC approach in [85] will be unavailable because the disturbances cannot be rejected accurately. In order to investigate further, it has been considered that uncertainties exist in such an exogenous model for the disturbance in (4.2). $\delta(t)$ is the additional disturbance signal resulting from the perturbations and uncertainties in the exogenous system (4.2), and satisfies 2-norm boundedness. In simulation, we select $\delta(t)$ as the random signal with upper 2-norm bound 1 and $d_1(t)$ as discrete wind gust model which is rooted in environment/wind of the aerospace

blockset in Simulink Library Browser of MATLAB 7.0. The mathematical representation of the discrete gust is

$$V_{wind} = \begin{cases} 0, & x < 0; \\ \frac{V_m}{2}(1 - \cos(\frac{\pi x}{d_m})), & 0 \le x \le d_m; \\ V_m, & x > d_m \end{cases}$$

where V_m is the gust amplitude, d_m is the gust length, x is the distance traveled, V_{wind}
is the resultant wind velocity in the body axis frame.

$$d_m = \begin{bmatrix} 120 & 120 & 80 \end{bmatrix}^T, \quad V_m = \begin{bmatrix} 3.5 & 3.5 & 3 \end{bmatrix}^T$$

are default block parameter in discrete wind gust model in Simulink Library Browser
of MATLAB 7.0.

4.5.1.1 The Case with Known Nonlinearity

Suppose that the nonlinear dynamics is denoted by

$$f_{01}(x(t),t) = sin(2\pi * 5t)x_2(t), f_{02}(x(t),t) = cos(2\pi * t)x_2(t).$$

The initial value of the state is taken to be

$$x(0) = \begin{bmatrix} 2 & -2 & 3 & 2 \end{bmatrix}^T$$

According to Remark 4.4, $\varepsilon = 0.01$. Select

$$\Theta = \begin{bmatrix} 20 & 0 \\ 0 & 20 \end{bmatrix}; \quad \Phi = \begin{bmatrix} 5 & 0 \\ 0 & 5 \end{bmatrix}; \quad r = 3/4; \quad q = 3; \quad p = 5;$$

$$\rho_1 = 5; \quad \rho_2 = 3; \quad Q_1 = \begin{bmatrix} 1 & 0 \\ 0 & 1 \end{bmatrix}; \quad \Lambda = \begin{bmatrix} 1 & 0 \\ 0 & 1 \end{bmatrix}$$

According to Theorems 4.1 and 4.2, it can be solved that

$$L = \begin{bmatrix} -1.0865 & -0.0458 & -0.0952 & -1.7697 \\ 0.1719 & -0.0280 & -0.0037 & -0.2830 \end{bmatrix} \times 10^3;$$

$$C_1 = \begin{bmatrix} 0.9146 & -0.2691 \\ -0.2149 & -0.9607 \end{bmatrix}; \quad C_2 = \begin{bmatrix} 1 & 0 \\ 0 & 1 \end{bmatrix};$$

$$C_3 = \begin{bmatrix} 0.9723 & -0.2287 \\ -0.2289 & -0.9702 \end{bmatrix}; \quad \gamma = 0.005$$

Figures 4.1-4.4 demonstrate the comparison of system performance between the DOBC scheme in [85] and the DOBPTSMC schemes. Figures 4.5 and 4.6 show the trajectory of DOBPTSM control and terminal sliding mode. Figure 4.7 and 4.8 illustrate the comparison of the system output response between DOBPTSMC and DOBC.

For the case with known nonlinearity, the effectiveness of the proposed DOBPTSMC scheme can be seen in Figures 4.1-4.4. The simulation results show that although there are exogenous disturbances in the system, the disturbance rejection and attenuation performance is improved and satisfactory system responses can be achieved with the proposed DOBPTSMC scheme compared with only the DOBC scheme. The simulation results shown in Figure 4.6 clearly demonstrate the finite time convergent property of TSM. It can be seen that the states of the system firstly reach TSM $S(t) = 0$ in the finite time $t_{r1} = 0.5775s$ in Figure 4.6, and converge to the neighborhood of $x(t) = 0$ in finite time $t_{r2} = 3.55s$ with DOBPTSMC. The range of the neighborhood of the equilibrium points to the simulation being consistent with that derived from the theoretical analysis. The simulation results shown in Figures 4.7 and 4.8 show that the output response is more satisfying with the proposed DOBPTSMC scheme than the pure DOBC scheme.

4.5.1.2 The Case with Unknown Nonlinearity

Different from Section 4.5.1.1, we suppose that $f_{01}(x(t),t)$, $f_{02}(x(t),t)$ are unknown in this section. Similar to [85], we assume

$$f_{01}(x(t),t) = r_1(t)x_2(t); \quad f_{02}(x(t),t) = r_2(t)x_1(t)$$

where $r_1(t)$, $r_2(t)$ are assumed to be random input with an upper bound 1, $f_{01}(x(t),t)$, $f_{02}(x(t),t)$ satisfy (4.3) and

$$\beta_{01} = 0, \; \beta_{11} = 1, \; \beta_{02} = 0, \; \beta_{12} = 1.$$

Suppose the initial value is

$$x(0) = \begin{bmatrix} 2 & -2 & 3 & 2 \end{bmatrix}^T.$$

According to Remark 4.4, $\varepsilon = 0.01$. Select

$$\Theta = \begin{bmatrix} 20 & 0 \\ 0 & 20 \end{bmatrix}; \; \Phi = \begin{bmatrix} 5 & 0 \\ 0 & 5 \end{bmatrix}; \; r = 3/4; \; q = 3; \; p = 5;$$

$$\rho_1 = 5; \; \rho_2 = 3; \; Q_1 = \begin{bmatrix} 1 & 0 \\ 0 & 1 \end{bmatrix}; \; \Lambda = \begin{bmatrix} 1 & 0 \\ 0 & 1 \end{bmatrix}$$

Based on Theorems 4.1 and 4.2, it can be solved that

$$L = \begin{bmatrix} -0.3141 & -0.0032 & -0.0000 & 0.3141 \\ -6.4501 & -0.0992 & -0.0000 & 6.4501 \end{bmatrix} \times 10^3;$$

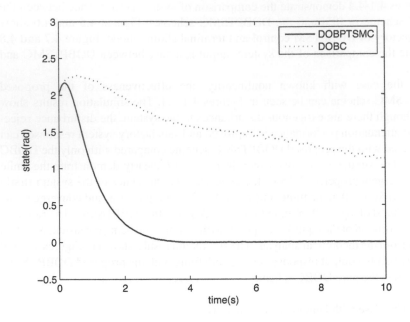

FIGURE 4.1 The comparison of system performance for the sideslip angle between DOBPTSMC and DOBC.

$$C1 = \begin{bmatrix} 0.9146 & -0.2691 \\ -0.2149 & -0.9607 \end{bmatrix}; \quad C_2 = \begin{bmatrix} 1 & 0 \\ 0 & 1 \end{bmatrix};$$

$$C3 = \begin{bmatrix} 0.9723 & -0.2287 \\ -0.2289 & -0.9702 \end{bmatrix}; \quad \gamma = 0.0048$$

Figures 4.9-4.12 show the comparison of system performance for the case with unknown nonlinearity between the DOBC scheme in [85] and the DOBPTSMC schemes. Figures 4.13 and 4.14 demonstrate the trajectory of DOBPTSM control and terminal sliding mode. Figures 4.15 and 4.16 illustrate the comparison of the system output response between DOBPTSMC and DOBC.

For the case with unknown nonlinearity, the proposed DOBPTSMC control strategy shown in Figure 4.13 can guarantee that the system states reach the sliding mode manifold $S(t) = 0$, and subsequently converge to the neighborhood of the equilibrium point $x(t) = 0$ in finite time. The simulation results shown in Figures 4.9-4.12 and 4.15 and 4.16 demonstrate that although there are exogenous disturbances in the system, the disturbance rejection and attenuation performance is improved and satisfactory system state responses and output responses can be achieved with the proposed DOBPTSMC scheme compared with the pure DOBC scheme.

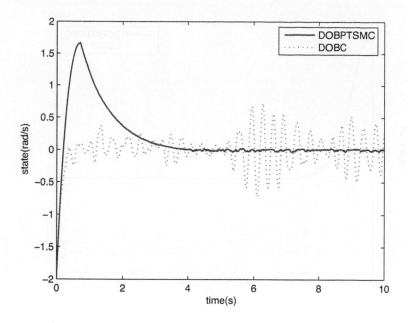

FIGURE 4.2   The comparison of system performance for the yaw rate between DOBPTSMC and DOBC.

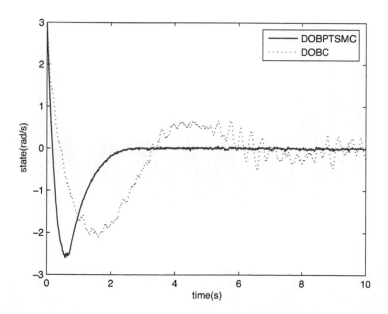

FIGURE 4.3   The comparison of system performance for the roll rate between DOBPTSMC and DOBC.

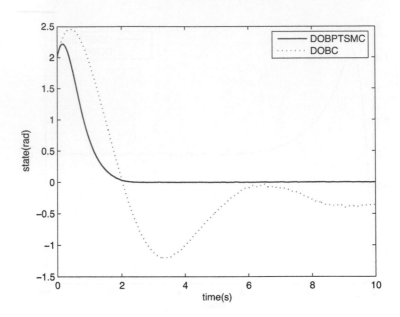

FIGURE 4.4 The comparison of system performance for the bank angle between DOBPTSMC and DOBC.

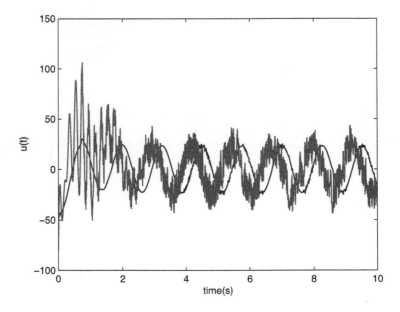

FIGURE 4.5 The trajectory of DOBPTSM control.

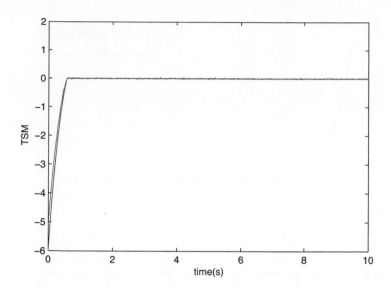

FIGURE 4.6   The trajectory of terminal sliding mode with DOBPTSMC.

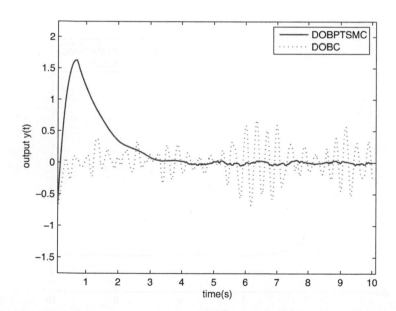

FIGURE 4.7   The comparison of the system output response between DOBPTSMC and DOBC y_1.

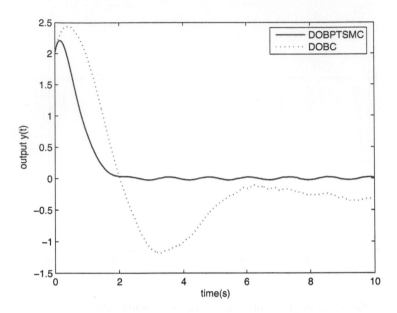

FIGURE 4.8   The comparison of the system output response between DOBPTSMC and DOBC y_2.

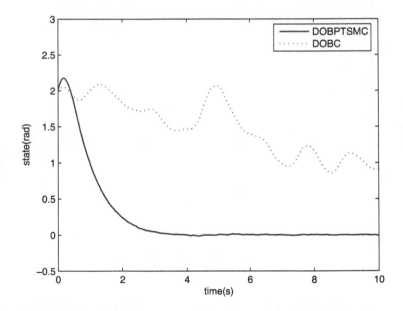

FIGURE 4.9   The comparison of system performance for the sideslip angle between DOBPTSMC and DOBC.

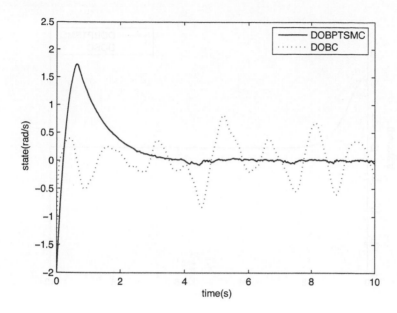

FIGURE 4.10   The comparison of system performance for the yaw rate between DOBPTSMC and DOBC.

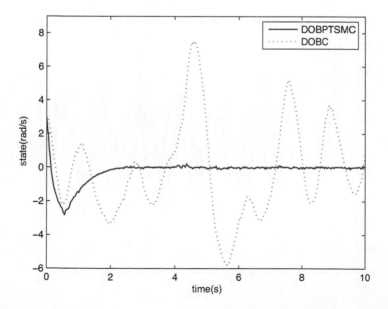

FIGURE 4.11   The comparison of system performance for the roll rate between DOBPTSMC and DOBC.

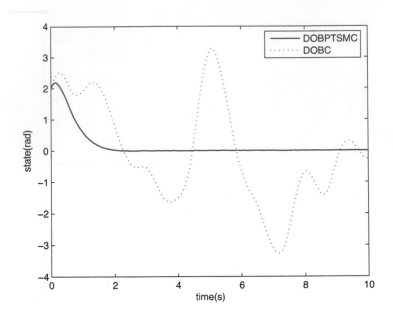

FIGURE 4.12 The comparison of system performance for the bank angle between DOBPTSMC and DOBC.

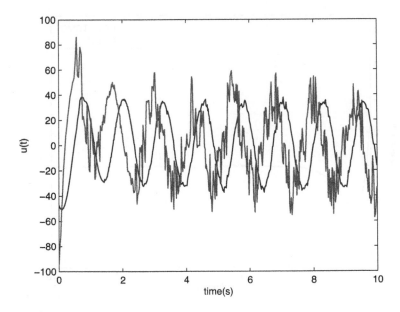

FIGURE 4.13 The trajectory of DOBPTSM control.

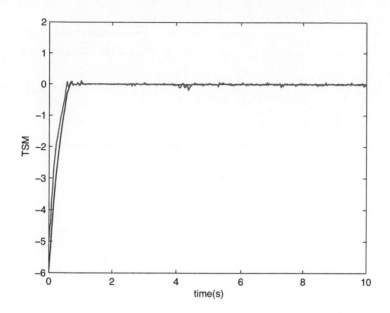

FIGURE 4.14   The trajectory of terminal sliding mode with DOBPTSMC.

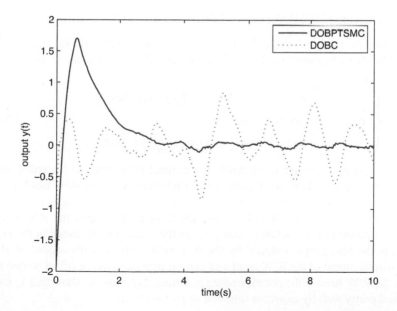

FIGURE 4.15   The comparison of the system output response between DOBPTSMC and DOBC y_1.

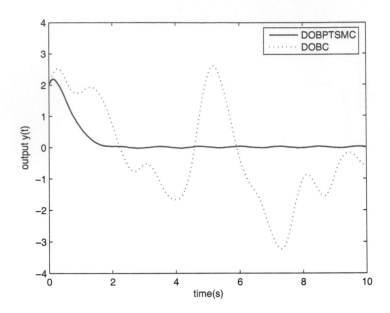

FIGURE 4.16   The comparison of the system output response between DOBPTSMC and DOBC y_2.

4.5.2   SIMULATION ON HARD DISK DRIVE ACTUATOR

A comprehensive hard disk drive (HDD) actuator model based on ([48] and [233]) which explicitly incorporates nonlinear dynamics and disturbances can be described as

$$
\begin{cases}
\ddot{z}(t) &= a[u(t) - T_c + d_0(t)] + \beta(t)\phi(t) + bd_1(t) \\
\dot{\phi}(t) &= \rho\phi(t) + \frac{1}{l}\dot{z}(t) \\
y(t) &= z(t) + F_{02}f_{02}(z(t),t) + D_0 d_0(t) + D_1 d_1(t)
\end{cases} \tag{4.52}
$$

where $u(t)$, $y(t)$ and $z(t)$ are respectively the control input voltage (in volts), the measurement output and the displacement (in micrometers) of the R/W head; $\beta(t)$ is nonlinear function satisfy $\beta(t) = \sigma + \beta_0(t)$, σ is an adjustable parameter. $\phi(t)$ is rotation angle of arm, $\rho < 0$ and l is the length of bracket from the R/W head to voice coil motor (VCM) actuator, and $d_1(t)$ is the friction disturbance in the H_2-norm. T_c is the bias torque induced by the flex cable, which is dependent on the position/displacement of the R/W head and can be measured as the offset torque to maintain the R/W head at the corresponding position. Experiments show that T_c can be matched pretty well by an arctan function of $z(t)$ in ([48]).

$$
T_c = \kappa \arctan(cz) \tag{4.53}
$$

where κ is a constant.

By using $x(t) = \begin{bmatrix} z^T(t) & \dot{z}^T(t) & \phi^T(t) \end{bmatrix}^T$, Equation (4.52) can be formulated as

$$\begin{cases} \dot{x}(t) &= G_0 x(t) + F_{01} f_{01}(x(t),t) + H_1 d_1(t) \\ &\quad + H_0[u(t) - T_c + d_0(t)] \\ y(t) &= C_0 x(t) + F_{02} f_{02}(x(t),t) + D_0 d_0(t) + D_1 d_1(t) \end{cases} \quad (4.54)$$

where

$$G_0 = \begin{bmatrix} 0 & 1 & 0 \\ 0 & 0 & \sigma \\ 0 & \frac{1}{l} & \rho \end{bmatrix}; \quad F_{01} = \begin{bmatrix} 0 \\ 1 \\ 0 \end{bmatrix};$$

$$H_0 = \begin{bmatrix} 0 \\ a \\ 0 \end{bmatrix}; \quad H_1 = \begin{bmatrix} 0 \\ b \\ 0 \end{bmatrix}; \quad C_0 = [1\ 0\ 0];$$

$$f_{01}(x(t),t) = \beta_0(t)x_3(t); \quad D_0 = 0.01; \quad D_1 = 0.01; \quad F_{02} = 0.01 \quad (4.55)$$

Denote $\tilde{u}(t) = u(t) - T_c$, then it is shown that

$$\begin{cases} \dot{x}(t) &= G_0 x(t) + F_{01} f_{01}(x(t),t) + H_1 d_1(t) \\ &\quad + H_0[\tilde{u}(t) + d_0(t)] \\ y(t) &= C_0 x(t) + F_{02} f_{02}(x(t),t) + D_0 d_0(t) + D_1 d_1(t) \end{cases} \quad (4.56)$$

According to [48], the coefficient parameter of HDD model are given by

$$a = 2.35 * 10^8; \ b = 3.8; \ c = 0.5886;$$

$$\sigma = 100; \ l = 1000; \ \rho = -100; \ \kappa = 0.02887;$$

$f_{01}(x(t),t)$, $f_{02}(x(t),t)$ represent nonlinear dynamics; $d_0(t)$ is rooted in repeated disturbance with known frequency and unknown phase in HDD actuator described by (4.2) with

$$W = \begin{bmatrix} 0 & 5600 \\ -5600 & 0 \end{bmatrix}; \quad V = \begin{bmatrix} 0.02 & 0 \end{bmatrix}; \quad H_2 = \begin{bmatrix} 0.01 \\ 0.01 \end{bmatrix}.$$

4.5.2.1 The Case with Known Nonlinearity

Suppose that the nonlinear dynamics is denoted by

$$f_{01}(x(t),t) = (100 + 50 * sin(t))x_3(t),$$

$$f_{02}(x(t),t) = 50 * cos(t)x_1(t).$$

The initial value of the state is taken to be $x(0) = \begin{bmatrix} 0.1 & -0.2 & 0.1 \end{bmatrix}^T$. According to Remark 4.4, $\varepsilon = 0.001$. Select

$$\Theta = \begin{bmatrix} 20 & 0 \\ 0 & 20 \end{bmatrix}; \quad r = 3/4; \quad q = 3; \quad p = 5;$$

$$\rho_1 = 9; \quad \rho_2 = 5; \quad Q_1 = \begin{bmatrix} 1 & 0 \\ 0 & 1 \end{bmatrix}; \quad \Lambda = \begin{bmatrix} 1 & 0 \\ 0 & 1 \end{bmatrix}$$

According to Theorems 4.1 and 4.2, it can be solved that

$$L = \begin{bmatrix} 0 & -0.7910 & 0 \\ 0 & -0.0026 & 0 \end{bmatrix} \times 10^{-5}; \quad \gamma = 0.05;$$

$$C_1 = \begin{bmatrix} 1, & 0.099 \end{bmatrix}; \quad C_2 = 1; \quad C_3 = \begin{bmatrix} 1, & -0.0010 \end{bmatrix}$$

Figures 4.17-4.19 demonstrate the comparison of system performance between the DOBC scheme in [85] and the DOBPTSMC schemes. Figures 4.20 and 4.21 show the trajectory of DOBPTSM control and terminal sliding mode. Figure 4.22 illustrates the comparison of the HDD system output response between DOBPTSMC and DOBC.

For the case with known nonlinearity, the effectiveness of the proposed DOBPTSMC scheme can be seen in Figures 4.17-4.19. The simulation results show that although there are exogenous disturbances in the system, the disturbance rejection and attenuation performance is improved and satisfactory system responses can be achieved with the proposed DOBPTSMC scheme compared with only DOBC scheme. The simulation results shown in Figure 4.21 clearly demonstrate the finite time convergent property of TSM. It can be seen that the states of system firstly reach TSM $S(t) = 0$ in finite time, and converge to the neighborhood of $x(t) = 0$ in finite time with DOBPTSMC. The simulation results shown in Figure 4.22 illustrate the output response under disturbances is more satisfying with the proposed DOBPTSMC scheme than the pure DOBC scheme.

4.5.2.2 The Case with Unknown Nonlinearity

Different from Section 4.5.2.1, we suppose that $f_{01}(x(t),t)$ and $f_{02}(x(t),t)$ are unknown in this section. Similar to [85], we assume

$$f_{01}(x(t),t) = r_1(t)x_3(t); \quad f_{02}(x(t),t) = r_2(t)x_1(t)$$

where $r_1(t)$, $r_2(t)$ are assumed to be random input with an upper bound 100, $f_{01}(x(t),t)$, $f_{02}(x(t),t)$ satisfy (4.3) and

$$\beta_{01} = 0, \ \beta_{11} = 100, \ \beta_{02} = 0, \ \beta_{12} = 100.$$

The initial value of the state is taken to be $x(0) = \begin{bmatrix} 0.1 & -0.2 & 0.1 \end{bmatrix}^T$. According to Remark 4.4, $\varepsilon = 0.001$. Select

$$\Theta = \begin{bmatrix} 20 & 0 \\ 0 & 20 \end{bmatrix}; \quad r = 3/4; \quad q = 3; \quad p = 5;$$

$$p_1 = 9; \quad p_2 = 5; \quad Q_1 = \begin{bmatrix} 1 & 0 \\ 0 & 1 \end{bmatrix}; \quad \Lambda = \begin{bmatrix} 1 & 0 \\ 0 & 1 \end{bmatrix}$$

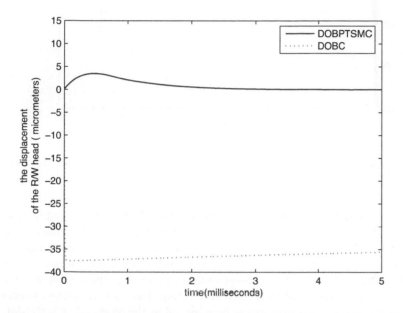

FIGURE 4.17   The comparison of system performance for the displacement of the R/W head between DOBPTSMC and DOBC.

According to Theorems 4.1 and 4.2, it can be solved that

$$L = \begin{bmatrix} 0 & -0.8125 & 0 \\ 0 & -0.0027 & 0 \end{bmatrix} \times 10^{-5}; \quad \gamma = 0.044;$$

$$C_1 = \begin{bmatrix} 1, & 0.099 \end{bmatrix}; \quad C_2 = 1; \quad C_3 = \begin{bmatrix} 1, & -0.001 \end{bmatrix}$$

Figures 4.23-4.25 show the comparison of system performance for the case with unknown nonlinearity between the DOBC scheme in [85] and the DOBPTSMC schemes. Figure 4.26 illustrates the comparison of HDD system output response between DOBPTSMC and DOBC. Figures 4.27 and 4.28 demonstrate the trajectory of DOBPTSM control and terminal sliding mode.

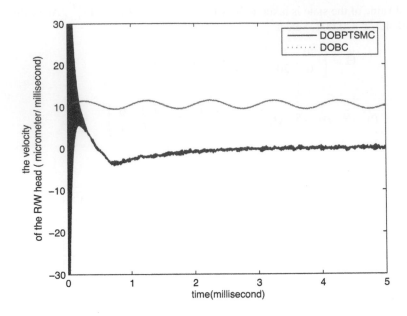

FIGURE 4.18   The comparison of system performance for the velocity of the R/W head between DOBPTSMC and DOBC.

For the case with unknown nonlinearity, the proposed DOBPTSMC control strategy shown in Figure 4.27 can guarantee that the system states reach the sliding mode manifold $S(t) = 0$, and subsequently converge to the neighborhood of the equilibrium point $x(t) = 0$ in finite time. The simulation results shown in Figures 4.22-4.26 demonstrate that although there are exogenous disturbances in the system, the disturbance rejection and attenuation performance is improved and satisfactory system state responses and output responses can be achieved by the proposed DOBPTSMC scheme compared with the pure DOBC scheme.

4.6   CONCLUSION

It has been well known that both TSM control and DOBC are effective robust control schemes against external disturbances and modeling perturbations. DOBC has been used to reject the influence of the disturbance with some known information. And TSM control can generally attenuate the effect of the norm-bounded unknown disturbances. In this chapter, these two schemes are combined to form a novel composite framework, which can be abbreviated as the DOB plus TSM (DOBPTSM) control. A class of MIMO continuous systems is considered, where the disturbance has been divided into two parts. One is supposed to be a norm-bounded vector. The other is described by an exogenous system with perturbations. With such a composite control scheme, the two different types of disturbance can be rejected and attenuated, respectively.

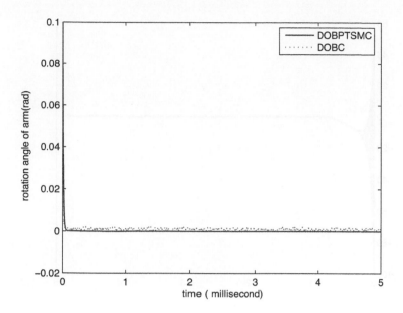

FIGURE 4.19   The comparison of system performance for the rotation of the R/W head between DOBPTSMC and DOBC.

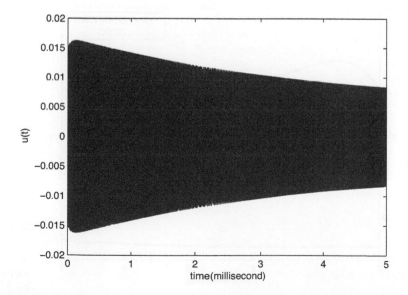

FIGURE 4.20   The trajectory of DOBPTSM control.

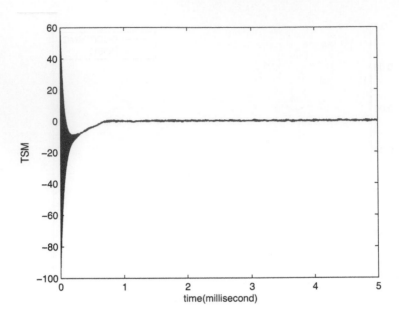

FIGURE 4.21   The trajectory of terminal sliding mode with DOBPTSMC.

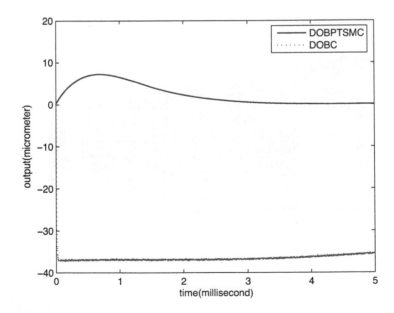

FIGURE 4.22   The trajectory of DOBPTSM control and terminal sliding mode.

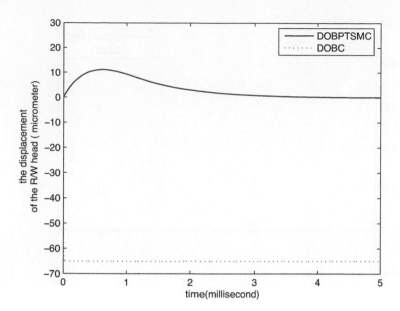

FIGURE 4.23   The comparison of system performance for the displacement of the R/W head with DOBPTSMC and DOBC.

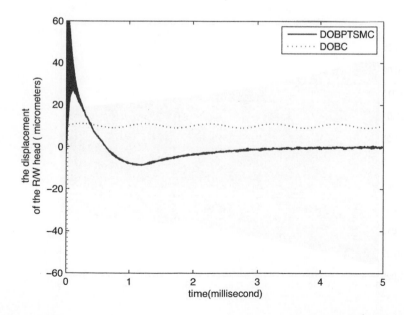

FIGURE 4.24   The comparison of system performance for the velocity of the R/W head between DOBPTSMC and DOBC.

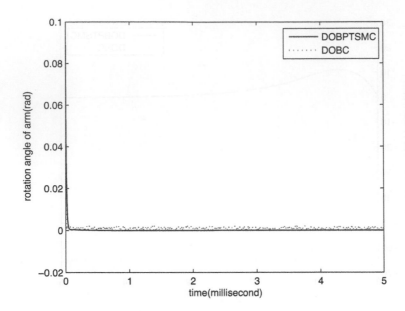

FIGURE 4.25 The comparison of system performance for rotation angle of arm between DOBPTSMC and DOBC.

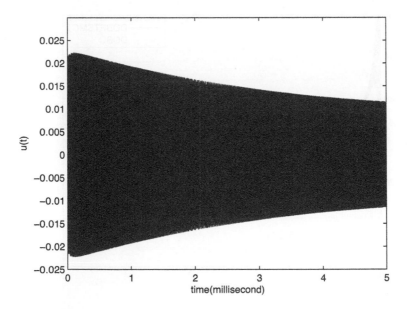

FIGURE 4.26 The trajectory of DOBPTSM control.

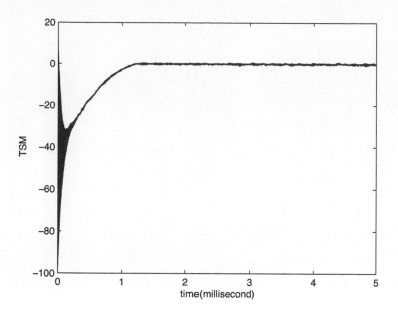

FIGURE 4.27   The trajectory of terminal sliding mode with DOBPTSMC.

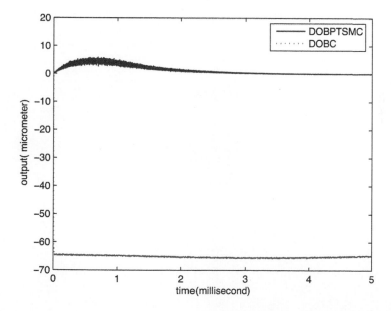

FIGURE 4.28   The trajectory of DOBPTSM control and terminal sliding mode.

5 Saturating Composite DOBC and H_∞ Control for Discrete Time Delay Systems

5.1 INTRODUCTION

Basically, control methods addressing disturbance attenuation can be classified according to the assumptions made on disturbances. An effective means of suppressing the effects of disturbance in a class of nonlinear systems is the application of disturbance observers since the late 1980s. The motivation is suggested by the fact that if the disturbances can be estimated, then control of the uncertain dynamic systems with disturbances may become easier. Later, the theory of disturbance observers was advanced in reference [195]. Dynamical systems with time delays have been one of the most active research fields, since time delays exist in many systems, such as hydraulic processes, chemical systems, and temperature processes, are often a primary source of instability and performance degradation. So it is crucial to design high-performance controllers for time delay systems and a great number of research results have been reported for time delay systems over the past few decades [66, 67, 84, 88, 140, 190, 239]. In addition, the study of systems subject to input saturation has been attracting the attention of researchers for a long time, in view of the universality of this issue in practical engineering. For example, the phenomenon of amplitude saturation in actuators, though frequently ignored, is due to inherent physical limitations of devices, which cannot be avoided. Moreover, failure in accounting for actuator saturation may lead to severe deterioration of closed loop system performance, even to instability. The stabilization of time delay systems with saturating control was investigated in [40, 166]. In [40], a linear dynamic output feedback compensator was designed. In [166], the global stabilization of observer-based linear constrained uncertain time delay systems was considered. However, disturbance observer based control (DOBC) with saturation has not been presented for time delay systems subject to multiple disturbances.

In this chapter, the stabilization problem for a class of time delay systems with nonlinear perturbation and disturbances is considered. A reduced-order observer is structured for estimation of the disturbance. A saturating composite DOBC and H_∞ control is proposed for discrete time delay systems so that the disturbance can be compensated for and attenuated. Simulations demonstrate the advantage of the proposed composite control scheme with saturation.

5.2 FORMULATION OF THE PROBLEM

The following discrete time delay system with nonlinear perturbation and two types of disturbances is considered

$$
\begin{cases}
x(k+1) &= G_0 x(k) + G_{0d} x(k-T) + F_{01} f_{01}(x(k), k) \\
&\quad + H_0 [u(k) + d_0(k)] + H_1 d_1(k) \\
y(k) &= C_0 x(k) + C_{0d} x(k-T) + F_{02} f_{02}(x(k), k)
\end{cases}
\tag{5.1}
$$

where $x(k)$ is the state, $y(k)$ is the measurement output. G_0, G_{0d}, H_0, H_1, C_0 and C_{0d} are the coefficient matrices. F_{01} and F_{02} are the corresponding weighting matrices. f_{01} and f_{02} are nonlinear functions which are supposed to satisfy bounded conditions described as Assumption 5.2. T is a known constant delay. $d_0(k)$ is supposed to be described by an exogenous system in Assumption 5.1, which can represent the constant and the harmonic noises, etc. $d_1(k)$ is the parametric uncertainties and the external disturbance in the H_2-norm.

Assumption 5.1 *The disturbance $d_0(k)$ in the control input path can be formulated by the following exogenous system*

$$
\begin{cases}
w(k+1) &= W w(k) + H_2 \delta(k) \\
d_0(k) &= V w(k)
\end{cases}
\tag{5.2}
$$

where W, H_2 and V are known matrices. $\delta(k)$ is the additional disturbance which results from the perturbations and uncertainties in the exogenous system.

Assumption 5.2 *For the known matrices $U_{0i}, i = 0, 1$, the nonlinear functions satisfy*

$$
\begin{cases}
\|f_{0i}(x_1(k), k) - f_{0i}(x_2(k), k)\| \le \|U_{0i}(x_1(k) - x_2(k))\|, \\
f_{0i}(0, k) = 0, \quad i = 1, 2
\end{cases}
\tag{5.3}
$$

Assumption 5.3 *(G_0, H_0) is controllable; $(W, H_0 V)$ is observable.*

Suppose that the system state is available. If $d_0(k)$ is considered as a part of the augmented state, then a reduced-order observer is needed. In this chapter, we construct reduced-order observers for $d_0(k)$ with known nonlinear dynamics and unknown nonlinear dynamics, respectively. Then we design a composite controller with saturation so that the disturbance can be rejected and the stability of the composite system can be guaranteed.

5.3 COMPOSITE CONTROL WITH SATURATION FOR THE CASE WITH KNOWN NONLINEARITY

In this section, we suppose that nonlinear functions f_{01} and f_{02} are given and Assumptions 5.1-5.3 hold.

5.3.1 DOBC WITH SATURATION

In this section, the disturbance observer is formulated as

$$\begin{cases} \hat{d}_0(k) & = & V\hat{w}(k) \\ \hat{w}(k) & = & v(k) - Lx(k) \\ v(k+1) & = & (W + LH_0V)(v(k) - Lx(k)) + L(G_0x(k) \\ & & + G_{0d}x(k-T) + F_{01}f_{01}(x(k),k) + H_0u(k)) \end{cases} \tag{5.4}$$

where $\hat{w}(k)$ is the estimation of $w(k)$, and $v(k)$ is the auxiliary vector as the state of the observer. The estimation error is denoted as

$$e_w(k) = w(k) - \hat{w}(k)$$

Based on (5.1), (5.2) and (5.4), it can be verified that

$$e_w(k+1) = (W + LH_0V)e_w(k) + H_2\delta(k) + LH_1d_1(k) \tag{5.5}$$

The objective of disturbance rejection can be achieved by designing the observer gain such that (5.5) satisfies the desired stability and robustness performance.

The DOB controller with saturation is designed as

$$u(k) = -\hat{d}_0(k) + \hat{u}(k), \quad \hat{u}(k) = sat(\Psi(k)), \quad \Psi(k) = 2Kx(k) \tag{5.6}$$

where

$$sat(\Psi(k)) = \begin{bmatrix} sat(\Psi_1(k)) & \cdots & sat(\Psi_m(k)) \end{bmatrix}^T \tag{5.7}$$

is a nonlinear function whose element is defined by $(i = 1,...m)$

$$sat(\Psi_i(k)) = \begin{cases} \underline{\Psi}_i, & \Psi_i(k) \leq \underline{\Psi}_i < 0 \\ \Psi_i(k), & \underline{\Psi}_i \leq \Psi_i(k) \leq \overline{\Psi}_i \\ \overline{\Psi}_i, & \Psi_i(k) \geq \overline{\Psi}_i > 0 \end{cases} \tag{5.8}$$

In the definition, $\underline{\Psi}_i$ and $\overline{\Psi}_i$ are given parameters based on the physical characteristics of the considered system.

Substituting (5.6) into Equation (5.1) yields the closed-loop system formulated by

$$\begin{aligned} x(k+1) & = & (G_0 + H_0K)x(k) + G_{0d}x(k-T) + F_{01}f_{01}(x(k),k) \\ & & + H_0Ve_w(k) + H_0\upsilon(k) + H_1d_1(k) \end{aligned} \tag{5.9}$$

where $\upsilon(k) = sat(2Kx(k)) - Kx(k)$ is an auxiliary vector. It is easy to verify that $\upsilon(k)$ satisfies

$$\upsilon^T(k)\upsilon(k) \leq x^T(k)K^T Kx(k)$$

Thus, the composite system combined (5.5) with (5.9) yields

$$
\begin{bmatrix} x(k+1) \\ e_w(k+1) \end{bmatrix} = \begin{bmatrix} G_0 + H_0 K & H_0 V \\ 0 & W + L H_0 V \end{bmatrix} \begin{bmatrix} x(k) \\ e_w(k) \end{bmatrix}
$$

$$
+ \begin{bmatrix} G_{0d} & 0 \\ 0 & 0 \end{bmatrix} \begin{bmatrix} x(k-T) \\ e_w(k-T) \end{bmatrix} + \begin{bmatrix} F_{01} \\ 0 \end{bmatrix} f_{01}(x(k),k)
$$

$$
+ \begin{bmatrix} H_0 \\ 0 \end{bmatrix} \upsilon(k) + \begin{bmatrix} H_1 & 0 \\ L H_1 & H_2 \end{bmatrix} \begin{bmatrix} d_1(k) \\ \delta(k) \end{bmatrix}
$$

And the concerned system can be formulated by

$$
\begin{cases} \bar{x}(k+1) &= G\bar{x}(k) + G_d \bar{x}(k-T) + F f(\bar{x}(k),k) \\ &\quad + E\upsilon(k) + Hd(k) \\ z(k) &= C\bar{x}(k) + C_d \bar{x}(k-T) + Dd(k) \end{cases} \tag{5.10}
$$

where

$$
\bar{x}(k) = \begin{bmatrix} x(k) \\ e_w(k) \end{bmatrix}; \quad \bar{x}(k-T) = \begin{bmatrix} x(k-T) \\ e_w(k-T) \end{bmatrix};
$$

$$
G = \begin{bmatrix} G_0 + H_0 K & H_0 V \\ 0 & W + L H_0 V \end{bmatrix}; \quad F = \begin{bmatrix} F_{01} \\ 0 \end{bmatrix};
$$

$$
G_d = \begin{bmatrix} G_{0d} & 0 \\ 0 & 0 \end{bmatrix}; \quad f(\bar{x}(k),k) = f_{01}(x(k),k);
$$

$$
E = \begin{bmatrix} H_0 \\ 0 \end{bmatrix}; \quad H = \begin{bmatrix} H_1 & 0 \\ L H_1 & H_2 \end{bmatrix}; \quad d(k) = \begin{bmatrix} d_1(k) \\ \delta(k) \end{bmatrix} \tag{5.11}
$$

For $f(\bar{x}(k),k)$, it can be seen that

$$
\|f(\bar{x}(k),k)\| \leq \|U_1 \bar{x}(k)\|
$$

where $U_1 = \begin{bmatrix} U_{01} & 0 \end{bmatrix}$. In (5.10), $z(k)$ is denoted as the reference output, and the weighting matrices are

$$
C = \begin{bmatrix} C_1 & C_2 \end{bmatrix}; \quad C_d = \begin{bmatrix} C_{d1} & C_{d2} \end{bmatrix}; \quad D = \begin{bmatrix} D_1 & D_2 \end{bmatrix}
$$

With the above formulation, the main objective is to construct an observer to estimate the disturbance $d_0(k)$, and then compute a composite DOBC and H_∞ controller with saturation such that (5.10) is stable and the performance of disturbance attenuation is satisfied. The new control scheme combines the classical DOBC with H_∞ control, which can be called the "DOBC plus H_∞ control," and abbreviated as DOBPH_∞C.

5.3.2 COMPOSITE DOBC AND H_∞ CONTROL

In this section, the main task is to design L and K such that (5.10) is asymptotically stable and satisfies the performance requirements for disturbance attenuation.

Theorem 5.1

For given parameters $\lambda_1 > 0$, $\lambda_2 > 0$, $\gamma > 0$, if there exist $Q_1 > 0$, $P_2 > 0$, $N_1 > 0$, $S_2 > 0$ and R_1, R_2 satisfying

$$
\left[
\begin{array}{cccccc}
-Q_1 & * & * & * & * & * \\
0 & -P_2+S_2 & * & * & * & * \\
0 & 0 & -N_1 & * & * & * \\
0 & 0 & 0 & -S_2 & * & * \\
0 & 0 & 0 & 0 & -\gamma^2 I & * \\
0 & 0 & 0 & 0 & 0 & M_{66} \\
G_0Q_1+H_0R_1 & H_0V & G_{0d}N_1 & 0 & M_{75} & M_{76} \\
0 & P_2W+R_2H_0V & 0 & 0 & M_{85} & 0 \\
C_1Q_1 & C_2 & C_{d1}N_1 & C_{d2} & M_{95} & 0 \\
M_{101} & 0 & 0 & 0 & 0 & 0 \\
Q_1 & 0 & 0 & 0 & 0 & 0
\end{array}
\right.
$$

$$
\left.
\begin{array}{ccccc}
* & * & * & * & * \\
* & * & * & * & * \\
* & * & * & * & * \\
* & * & * & * & * \\
* & * & * & * & * \\
* & * & * & * & * \\
-Q_1 & * & * & * & * \\
0 & -P_2 & * & * & * \\
0 & 0 & -I & * & * \\
0 & 0 & 0 & M_{66} & * \\
0 & 0 & 0 & 0 & -N_1
\end{array}
\right] < 0 \qquad (5.12)
$$

where

$$
M_{66} = \begin{bmatrix} -\lambda_1^2 I & 0 \\ 0 & -\lambda_2^2 I \end{bmatrix}, M_{101} = \begin{bmatrix} U_{01}Q_1 \\ KQ_1 \end{bmatrix}, M_{75} = \begin{bmatrix} H_1 & 0 \end{bmatrix},
$$

$$
M_{85} = \begin{bmatrix} R_2H_1 & P_2H_2 \end{bmatrix}, M_{95} = \begin{bmatrix} D_1 & D_2 \end{bmatrix}, M_{76} = \begin{bmatrix} F_{01} & H_0 \end{bmatrix}.
$$

then composite system (5.10) under saturating DOBC law (5.6) with gain $K = R_1Q_1^{-1}$ and observer (5.4) with gain $L = P_2^{-1}R_2$ is robustly asymptotically stable and satisfies $||z(k)||_2 < \gamma||d(k)||_2$. ∎

Proof. Consider the following Lyapunov function

$$V(k) = \bar{x}^T(k)P\bar{x}(k) + \lambda_1^2 \sum_{i=1}^{k-1}[||U_1\bar{x}(i)||^2 - ||f(\bar{x}(i),i)||^2]$$

$$+ \lambda_2^2 \sum_{i=1}^{k-1}[||K\bar{x}(i)||^2 - ||v(k)||^2] + \sum_{i=k-T}^{k-1} \bar{x}^T(k)S\bar{x}(k) \qquad (5.13)$$

Along with system (5.10) in the absence of the disturbance $d(k)$, it can be verified that

$$\Delta V(k) = V(k+1) - V(k)$$

$$= \begin{bmatrix} \bar{x}(k)^T & \bar{x}(k-T)^T & f(\bar{x}(k),k)^T & v(k)^T \end{bmatrix} M_1$$

$$\begin{bmatrix} \bar{x}(k)^T & \bar{x}(k-T)^T & f(\bar{x}(k),k)^T & v(k)^T \end{bmatrix}^T$$

where

$$M_1 = \begin{bmatrix} \Pi_1 & G^T PG_d & G^T PF & G^T PE \\ G_d^T PG & G_d^T PG_d - S & G_d^T PF & G_d^T PE \\ F^T PG & F^T PG_d & F^T PF - \lambda_1^2 I & F^T PE \\ E^T PG & E^T PG_d & E^T PF & E^T PE - \lambda_2^2 I \end{bmatrix}$$

where

$$\Pi_1 = G^T PG - P + S + \lambda_1^2 U_1^T U_1 + \lambda_2^2 K^T K.$$

It is noted that $M_1 < 0$ implies that the stability of the system in the absence of $d(k)$.

Denote an auxiliary function as

$$J_1 = \sum_{i=0}^{\infty}[z(i)^T z(i) - \gamma^2 d(i)^T d(i) + \Delta V(i)] \qquad (5.14)$$

Under the zero initial condition, it can be seen that

$$J_1 \geq \sum_{i=0}^{\infty}[z(i)^T z(i) - \gamma^2 d(i)^T d(i)]$$

Along with system (5.10), it can be verified that

$$z(k)^T z(k) - \gamma^2 d(k)^T d(k) + \Delta V(k) = \begin{bmatrix} \bar{x}(k) \\ \bar{x}(k-T) \\ d(k) \\ f(\bar{x}(k),k) \\ v(k) \end{bmatrix}^T M_2 \begin{bmatrix} \bar{x}(k) \\ \bar{x}(k-T) \\ d(k) \\ f(\bar{x}(k),k) \\ v(k) \end{bmatrix}$$

where

$$M_2 = \begin{bmatrix} \Pi_1 + C^T C & * \\ G_d^T PG + C_d^T C & G_d^T PG_d - S + C_d^T C_d \\ H^T PG + D^T C & H^T PG_d + D^T C_d \\ F^T PG & F^T PG_d \\ E^T PG & E^T PG_d \end{bmatrix}$$

$$\begin{bmatrix} * & & & & & * & & & & * & & & & * \\ & * & & & & & & * & & & & & * \\ H^TPH - \gamma^2 I + D^TD & & & & & * & & & & * \\ F^TPH & & & F^TPF - \lambda_1^2 I & & & * \\ E^TPF & & & E^TPF & & & E^TPE - \lambda_2^2 I \end{bmatrix}$$

It is shown that $J_1 < 0$ holds if $M_2 < 0$. It can be seen that $M_2 < 0 \Leftrightarrow M_3 < 0$ based on the Schur complement formula. Where

$$M_3 = \begin{bmatrix}
-P+S & 0 & 0 & 0 & 0 & G^TP & C^T & U_1^T & K^T \\
0 & -S & 0 & 0 & 0 & G_d^TP & C_d^T & 0 & 0 \\
0 & 0 & -\gamma^2 I & 0 & 0 & H^TP & D^T & 0 & 0 \\
0 & 0 & 0 & -\lambda_1^2 I & 0 & F^TP & 0 & 0 & 0 \\
0 & 0 & 0 & 0 & -\lambda_2^2 I & E^TP & 0 & 0 & 0 \\
PG & PG_d & PH & PF & PE & -P & 0 & 0 & 0 \\
C & C_d & D & 0 & 0 & 0 & -I & 0 & 0 \\
U_1 & 0 & 0 & 0 & 0 & 0 & 0 & -\lambda_1^{-2} I & 0 \\
K & 0 & 0 & 0 & 0 & 0 & 0 & 0 & -\lambda_2^{-2} I
\end{bmatrix} < 0 \qquad (5.15)$$

Thus, it is shown that $J_1 < 0$ holds under $M_3 < 0$, which further leads to $||z||_2 < \gamma ||d||_2$. Moreover, it is obvious that condition $M_3 < 0$ guarantees $M_1 < 0$. So for given parameter $\lambda_1 > 0, \lambda_1 > 0$ and $\gamma > 0$, there exists $P > 0$, $S > 0$ satisfying $M_3 < 0$, then system (5.10) is stable in the absence of $d(k)$ and satisfies $||z||_2 < \gamma ||d||_2$.

The next step is to show that for the composite system (5.10), $M_3 < 0$ can be guaranteed when inequality (5.12) holds. Firstly we denote

$$P = \begin{bmatrix} P_1 & 0 \\ 0 & P_2 \end{bmatrix} = \begin{bmatrix} Q_1^{-1} & 0 \\ 0 & P_2 \end{bmatrix} > 0,$$

$$S = \begin{bmatrix} S_1 & 0 \\ 0 & S_2 \end{bmatrix} = \begin{bmatrix} E_1^{-1} & 0 \\ 0 & S_2 \end{bmatrix} > 0 \qquad (5.16)$$

Substituting (5.11) and (5.16) into (5.15), yields M_4 shown by

$$M_4 = \begin{bmatrix} -P_1+S_1 & * & * & * & * \\ 0 & -P_2+S_2 & * & * & * \\ 0 & 0 & -S_1 & * & * \\ 0 & 0 & 0 & -S_2 & * \\ 0 & 0 & 0 & 0 & -\gamma^2 I \\ 0 & 0 & 0 & 0 & 0 \\ P_1G_0+P_1H_0K & P_1H_0V & P_1G_{0d} & 0 & P_1M_{75} \\ 0 & P_2W+P_2LH_0V & 0 & 0 & \Pi_{85} \\ C_1 & C_2 & C_{d1} & C_{d2} & M_{95} \\ \Pi_{101} & 0 & 0 & 0 & 0 \end{bmatrix}$$

$$\begin{bmatrix} * & * & * & * & * \\ * & * & * & * & * \\ * & * & * & * & * \\ * & * & * & * & * \\ * & * & * & * & * \\ M_{66} & * & * & * & * \\ P_1M_{76} & -P_1 & * & * & * \\ 0 & 0 & -P_2 & * & * \\ 0 & 0 & 0 & -I & * \\ 0 & 0 & 0 & 0 & M_{66} \end{bmatrix} < 0$$

where

$$\Pi_{85} = \begin{bmatrix} P_2LH_1 & P_2H_2 \end{bmatrix}, \Pi_{101} = \begin{bmatrix} U_{01} \\ K \end{bmatrix}$$

Furthermore, it can be shown that $M_5 = \Theta^T M_4 \Theta$, and

$$M_5 = \begin{bmatrix} -Q_1+Q_1S_1Q_1 & * & * & * & * \\ 0 & -P_2+S_2 & * & * & * \\ 0 & 0 & -S_1 & * & * \\ 0 & 0 & 0 & -S_2 & * \\ 0 & 0 & 0 & 0 & -\gamma^2 I \\ 0 & 0 & 0 & 0 & 0 \\ G_0Q_1+H_0KQ_1 & H_0V & G_{0d} & 0 & M_{75} \\ 0 & P_2W+P_2LH_0V & 0 & 0 & \Pi_{85} \\ C_1Q_1 & C_2 & C_{d1} & C_{d2} & M_{95} \\ M_{101} & 0 & 0 & 0 & 0 \end{bmatrix}$$

$$
\begin{bmatrix}
* & * & * & * & * \\
* & * & * & * & * \\
* & * & * & * & * \\
* & * & * & * & * \\
* & * & * & * & * \\
M_{66} & * & * & * & * \\
M_{76} & -Q_1 & * & * & * \\
0 & 0 & -P_2 & * & * \\
0 & 0 & 0 & -I & * \\
0 & 0 & 0 & 0 & M_{66}
\end{bmatrix} < 0
$$

where $\Theta = diag\{Q_1, I, I, I, I, I, I, Q_1, I, I, I, I\}$ and $Q_1 = P_1^{-1}$. Based on Schur complement, $M_5 < 0$ is equivalent to $M_6 < 0$, where

$$
M_6 = \begin{bmatrix}
-Q_1 & * & * & * & * \\
0 & -P_2 + S_2 & * & * & * \\
0 & 0 & -S_1 & * & * \\
0 & 0 & 0 & -S_2 & * \\
0 & 0 & 0 & 0 & -\gamma^2 I \\
0 & 0 & 0 & 0 & 0 \\
G_0 Q_1 + H_0 K Q_1 & H_0 V & G_{0d} & 0 & M_{75} \\
0 & P_2 W + P_2 L H_0 V & 0 & 0 & \Pi_{85} \\
C_1 Q_1 & C_2 & C_{d1} & C_{d2} & M_{95} \\
M_{101} & 0 & 0 & 0 & 0 \\
Q_1 & 0 & 0 & 0 & 0
\end{bmatrix}
$$

$$
\begin{bmatrix}
* & * & * & * & * & * \\
* & * & * & * & * & * \\
* & * & * & * & * & * \\
* & * & * & * & * & * \\
* & * & * & * & * & * \\
M_{66} & * & * & * & * & * \\
M_{76} & -Q_1 & * & * & * & * \\
0 & 0 & -P_2 & * & * & * \\
0 & 0 & 0 & -I & * & * \\
0 & 0 & 0 & 0 & M_{66} & * \\
0 & 0 & 0 & 0 & 0 & -S_1^{-1}
\end{bmatrix} < 0
$$

By defining $R_1 = KQ_1$, $R_2 = P_2 L$ and $N_1 = S_1^{-1}$, M_6 is pre-multiplied and post-multiplied simultaneously by $diag\{I, I, N_1, I, I, I, I, I, I, I, I, I, I\}$, then is equivalent to M_7, where

$$M_7 = \begin{bmatrix}
-Q_1 & * & * & * & * & * & * & * & * & * & * \\
0 & -P_2+S_2 & * & * & * & * & * & * & * & * & * \\
0 & 0 & -N_1 & * & * & * & * & * & * & * & * \\
0 & 0 & 0 & -S_2 & * & * & * & * & * & * & * \\
0 & 0 & 0 & 0 & -\gamma^2 I & * & * & * & * & * & * \\
0 & 0 & 0 & 0 & 0 & M_{66} & * & * & * & * & * \\
G_0Q_1+H_0R_1 & H_0V & G_{0d}N_1 & 0 & M_{75} & M_{76} & -Q_1 & * & * & * & * \\
0 & P_2W+R_2H_0V & 0 & 0 & \Pi_{85} & 0 & 0 & -P_2 & * & * & * \\
C_1Q_1 & C_2 & C_{d1}N_1 & C_{d2} & M_{95} & 0 & 0 & 0 & -I & * & * \\
M_{101} & 0 & 0 & 0 & 0 & 0 & 0 & 0 & 0 & M_{66} & * \\
Q_1 & 0 & 0 & 0 & 0 & 0 & 0 & 0 & 0 & 0 & -N_1
\end{bmatrix} < 0$$

So (5.12) can be obtained; the proof is completed. $\qquad\square$

5.4 COMPOSITE CONTROL WITH SATURATION FOR THE CASE WITH UNKNOWN NONLINEARITY

In this section, we also suppose Assumptions 5.1-5.3 hold, but nonlinear dynamics f_{01} and f_{02} are unknown. Different from Section 5.3, f_{01} are unavailable in disturbance observer design.

5.4.1 DOBC WITH SATURATION

In this section, the disturbance observer is formulated as

$$\begin{cases}
\hat{d}_0(k) &=& V\hat{w}(k) \\
\hat{w}(k) &=& v(k) - Lx(k) \\
v(k+1) &=& (W+LH_0V)(v(k) - Lx(k)) + L[G_0x(k) \\
&& + G_{0d}x(k-T) + H_0u(k)]
\end{cases} \qquad (5.17)$$

Compared with (5.5), the estimation error $e_w(k) = w(k) - \hat{w}(k)$ satisfies

$$e_w(k+1) = (W + LH_0V)e_w(k) + LF_{01}f_{01} + H_2\delta(k) + LH_1d_1(k) \tag{5.18}$$

The DOB controller with saturation is formulated as

$$u(k) = -\hat{d}_0(k) + \hat{u}(k), \quad \hat{u}(k) = sat(\Psi(k)), \quad \Psi(k) = 2Kx \tag{5.19}$$

Combing (5.19) with (5.1), the closed-loop system is described as

$$x(k+1) = (G_0 + H_0K)x(k) + G_{0d}x(k-T) + F_{01}f_{01}(x(k),k)$$

$$+ H_0Ve_w(k) + H_0v(k) + H_1d_1(k) \tag{5.20}$$

where $v(k) = sat(2Kx(k)) - Kx(k)$ is an auxiliary vector. Thus, the composite system combined (5.18) with (5.20) is given by

$$\begin{bmatrix} x(k+1) \\ e_w(k+1) \end{bmatrix} = \begin{bmatrix} G_0 + H_0K & H_0V \\ 0 & W + LH_0V \end{bmatrix} \begin{bmatrix} x(k) \\ e_w(k) \end{bmatrix}$$

$$+ \begin{bmatrix} G_{0d} & 0 \\ 0 & 0 \end{bmatrix} \begin{bmatrix} x(k-T) \\ e_w(k-T) \end{bmatrix} + \begin{bmatrix} F_{01} \\ LF_{01} \end{bmatrix} f_{01}(x(k),k)$$

$$+ \begin{bmatrix} H_0 \\ 0 \end{bmatrix} v(k) + \begin{bmatrix} H_1 & 0 \\ LH_1 & H_2 \end{bmatrix} \begin{bmatrix} d_1(k) \\ \delta(k) \end{bmatrix}$$

And the concerned system can be formulated by

$$\bar{x}(k+1) = G\bar{x}(k) + G_d\bar{x}(k-T) + Ff(\bar{x}(k),k)$$

$$+ Ev(k) + Hd(k)$$

$$z(k) = C\bar{x}(k) + C_d\bar{x}(k-T) + Dd(k) \tag{5.21}$$

where

$$\bar{x}(k) = \begin{bmatrix} x(k) \\ e_w(k) \end{bmatrix}; \quad \bar{x}(k-T) = \begin{bmatrix} x(k-T) \\ e_w(k-T) \end{bmatrix};$$

$$G = \begin{bmatrix} G_0 + H_0K & H_0V \\ 0 & W + LH_0V \end{bmatrix}; \quad F = \begin{bmatrix} F_{01} \\ LF_{01} \end{bmatrix};$$

$$G_d = \begin{bmatrix} G_{0d} & 0 \\ 0 & 0 \end{bmatrix}; \quad f(\bar{x}(k),k) = f_{01}(x(k),k);$$

$$E = \begin{bmatrix} H_0 \\ 0 \end{bmatrix}; \quad H = \begin{bmatrix} H_1 & 0 \\ LH_1 & H_2 \end{bmatrix}; \quad d(k) = \begin{bmatrix} d_1(k) \\ \delta(k) \end{bmatrix}$$

In (5.21), $z(k)$ is denoted as the reference output, where the weighting matrices are

$$C = \begin{bmatrix} C_1 & C_2 \end{bmatrix}, \quad C_d = \begin{bmatrix} C_{d1} & C_{d2} \end{bmatrix}, \quad D = \begin{bmatrix} D_1 & D_2 \end{bmatrix}$$

5.4.2 COMPOSITE DOBC AND H_∞ CONTROL

Similarly to Section 5.3.2, in this section, our aim is to design L and K in order that (5.21) is asymptotically stable and satisfies the performance requirements for disturbance attenuation.

Theorem 5.2

For given parameters $\lambda_1 > 0$, $\lambda_2 > 0$, $\gamma > 0$, if there exist $Q_1 > 0$, $P_2 > 0$, $N_1 > 0$, $S_2 > 0$ and R_1, R_2 satisfying

$$
\begin{bmatrix}
-Q_1 & * & * & * & * & * \\
0 & -P_2 + S_2 & * & * & * & * \\
0 & 0 & -N_1 & * & * & * \\
0 & 0 & 0 & -S_2 & * & * \\
0 & 0 & 0 & 0 & -\gamma^2 I & * \\
0 & 0 & 0 & 0 & 0 & M_{66} \\
G_0 Q_1 + H_0 R_1 & H_0 V & G_{0d} N_1 & 0 & M_{75} & M_{76} \\
0 & P_2 W + R_2 H_0 V & 0 & 0 & M_{85} & M_{86} \\
C_1 Q_1 & C_2 & C_{d1} N_1 & C_{d2} & M_{95} & 0 \\
M_{101} & 0 & 0 & 0 & 0 & 0 \\
Q_1 & 0 & 0 & 0 & 0 & 0
\end{bmatrix}
$$

$$
\left.
\begin{bmatrix}
* & * & * & * & * \\
* & * & * & * & * \\
* & * & * & * & * \\
* & * & * & * & * \\
* & * & * & * & * \\
* & * & * & * & * \\
-Q_1 & * & * & * & * \\
0 & -P_2 & * & * & * \\
0 & 0 & -I & * & * \\
0 & 0 & 0 & M_{66} & * \\
0 & 0 & 0 & 0 & -N_1
\end{bmatrix}
\right] < 0 \qquad (5.22)
$$

where

$$
M_{86} = \begin{bmatrix} R_2 F_{01} & 0 \end{bmatrix}.
$$

then composite system (5.21) under saturating DOBC law (5.19) with gain $K = R_1 Q_1^{-1}$ and observer (5.17) with gain $L = P_2^{-1} R_2$ is robustly asymptotically stable and satisfies $\|z(k)\|_2 < \gamma \|d(k)\|_2$. ∎

Proof. The proof procedure can be given similarly to that of the proof for Theorem 5.1. □

Remark 5.1 For discrete time delay systems, many novel control strategies have been reported in [66, 140, 190, 239]. If the ideas could be utilized for solving the problems formulated in this chapter, some new results could be given.

5.5 SIMULATION ON A4D AIRCRAFT

In this section, it will be shown that robustness can be further improved for the discrete time delay models under the proposed composite control scheme with saturation. The longitudinal dynamics of A4D aircraft at a flight condition of 15,000-foot altitude and Mach 0.9 in [85] can be given by

$$x(k+1) = G_0 x(k) + G_{0d} x(k-T) + F_{01} f_{01}(x(k), k)$$
$$+ H_0 [u(k) + d_0(k)] + H_1 d_1(k)$$

where

$$x(k) = [\ x_1(k)\quad x_2(k)\quad x_3(k)\quad x_4(k)\ ]^T,$$

$x_1(k)$ is the forward velocity, $x_2(k)$ is the angle of attack, $x_3(k)$ is the pitching velocity, $x_4(k)$ is the pitching angle, $u(k)$ is elevator deflection and

$$G_0 = \begin{bmatrix} 0.9994 & 0.3209 & 0.0032 & 0.3219 \\ 0 & 0.9837 & 0.0098 & 0 \\ -0.0001 & -0.3397 & 0.9708 & 0 \\ 0 & -0.0017 & 0.0099 & 1 \end{bmatrix};$$

$$G_{0d} = \begin{bmatrix} 0.09 & 0.03 & 0.003 & 1.03 \\ 0.002 & 0.04 & 0.07 & -0.0034 \\ -0.004 & -0.04 & 0.03 & -0.001 \\ -0.0004 & -0.014 & 0.008 & 0.03 \end{bmatrix};$$

$$H_0 = \begin{bmatrix} -0.0005 \\ -0.0027 \\ -0.3329 \\ -0.0017 \end{bmatrix}; \quad F_{01} = \begin{bmatrix} 0 \\ 0 \\ 1 \\ 0 \end{bmatrix}; \quad H_1 = \begin{bmatrix} 0.010 \\ 0 \\ -0.038 \\ 0.010 \end{bmatrix};$$

$$U_{01} = \begin{bmatrix} 0 & 0 & 0 & 0 \\ 0 & 1 & 0 & 0 \\ 0 & 0 & 0 & 0 \\ 0 & 0 & 0 & 0 \end{bmatrix}; \quad T = 1$$

$z(k)$ is denoted as the reference output in (5.10), where

$$C_1 = [\ 1\quad 0\quad 0\quad 0\ ]; C_2 = [\ 0\quad 0\ ]; C_{d1} = [\ 1\quad 0\quad 0\quad 0\ ];$$

$$C_{d2} = [\; 0 \quad 0 \;]; D_1 = D_2 = 0$$

The exogenous system matrices for $d_0(k)$ are given by

$$W = \begin{bmatrix} 0 & 5 \\ -5 & 0 \end{bmatrix}; \; V = [\; 25 \quad 0 \;]; \; H_2 = \begin{bmatrix} 0.022 \\ 0.015 \end{bmatrix}$$

In simulation, we select $\delta(k)$ as the unknown signals with upper 2-norm bound 1 and $d_1(k)$ as discrete wind gust model. The mathematical representation of the discrete gust is

$$V_{wind} = \begin{cases} 0, & x < 0; \\ \frac{V_m}{2}(1 - \cos(\frac{\pi x}{d_m})), & 0 \le x \le d_m; \\ V_m, & x > d_m \end{cases}$$

where V_m is the gust amplitude, d_m is the gust length, x is the distance traveled, and V_{wind} is the resultant wind velocity in the body axis frame.

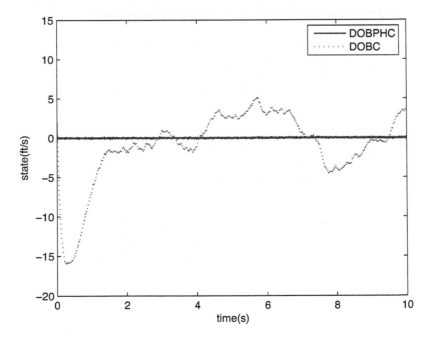

FIGURE 5.1 The comparison of system performance for the forward velocity with DOBPH_∞C and DOBC.

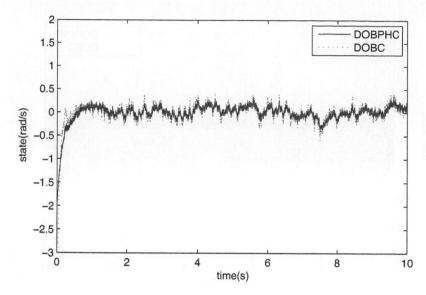

FIGURE 5.2   The comparison of system performance for angle of attack with DOBPH_∞C and DOBC.

5.5.1   THE CASE WITH KNOWN NONLINEARITY

Suppose that the nonlinear dynamics is denoted by

$$f_{01}(x(k),k) = sin(2\pi * 5k)x_2(k), \quad \|f_{01}(x(k),k)\| \le \|U_{01}x(k)\|,$$

where $U_{01} = diag\{0\ 1\ 0\ 0\}$. We select $\lambda_1 = \lambda_2 = 0.1$, $\gamma = 0.1$. The initial value of the state is taken to be

$$x(0) = \begin{bmatrix} 2 & -2 & 3 & 2 \end{bmatrix}^T.$$

Based on Theorem 5.1, it can be solved that

$$K = \begin{bmatrix} 115.8397 & 117.1857 & 3.6670 & 120.5320 \end{bmatrix};$$

$$L = \begin{bmatrix} -170.2507 & -33.6682 & 0.0016 & 170.2568 \\ -116.1210 & -22.9570 & -0.0206 & 116.0427 \end{bmatrix}$$

For the case with known nonlinearity, Figures 5.1-5.4 demonstrate the comparison of system performance with the DOBC scheme in [85] and the proposed DOBPH_∞C scheme with saturation. Figures 5.5-5.8 show the comparison of system performance with the H_∞C scheme and the DOBPH_∞C scheme with saturation.

The effectiveness of the proposed DOBPH_∞C scheme with saturation can be seen in Figures 5.1-5.8. The simulation results show that although there are exogenous disturbances in discrete time delay systems, the disturbance rejection performance

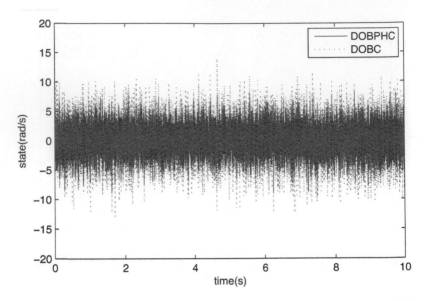

FIGURE 5.3 The comparison of system performance for the pitching velocity with DOBPH_∞C and DOBC.

is further improved and more satisfying system responses can be achieved by the proposed DOBPH_∞C scheme with saturation compared with only DOBC scheme or only H_∞C scheme.

5.5.2 THE CASE WITH UNKNOWN NONLINEARITY

Different from Section 5.5.1, here we suppose that $f_{01}(x(k),k)$ is unknown. In simulation, we assume

$$f_{01}(x(k),k) = r(k)x_2(k)$$

where $r(k)$ is assumed to be a random input with an upper bound 1. We select

$$\lambda_1 = \lambda_2 = 0.1, \ \gamma = 1.$$

Suppose the initial value is

$$x(0) = \begin{bmatrix} 2 & -2 & 3 & 2 \end{bmatrix}^T.$$

Based on Theorem 5.2, it can be obtained that

$$K = \begin{bmatrix} 113.5654 & 115.8527 & 3.6554 & 119.1904 \end{bmatrix};$$

$$L = \begin{bmatrix} -7.2173 & 34.7328 & 0.0015 & 7.2232 \\ -4.4129 & 21.2560 & 0.0006 & 4.4154 \end{bmatrix}$$

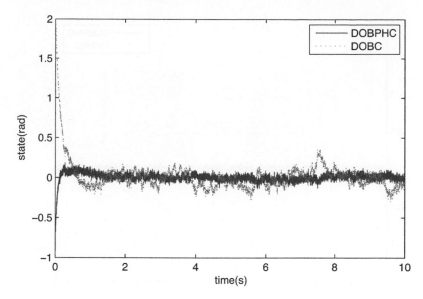

FIGURE 5.4 The comparison of system performance for the pitching angle with DOBPH_∞C and DOBC.

For the case with unknown nonlinearity, Figures 5.9-5.12 demonstrate the comparison of system performance with the DOBC scheme and the DOBPH_∞C scheme with saturation. Figures 5.13-5.16 show the comparison of system performance with the H_∞C scheme and the DOBPH_∞C scheme with saturation.

The effectiveness of the proposed DOBPH_∞C scheme with saturation is shown in Figures 5.9-5.16. The simulation results show that although there are exogenous disturbances in discrete time delay systems, system robustness performance achieved by the proposed DOBPH_∞C scheme with saturation is more satisfying than that by only DOBC scheme or only H_∞C scheme.

5.6 CONCLUSION

A class of discrete time delay systems with nonlinearity and disturbances are considered in this chapter, where the disturbance has been divided into two parts. One is supposed to be a norm-bounded vector; the other is described by an exogenous system with perturbations. A reduced-order observer is structured for the estimation of the disturbance. With saturating DOBPH_∞C scheme, the first type of disturbances can be estimated and rejected by DOBC, and the second one can be attenuated by H_∞ control. Simultaneously, the desired dynamic performances for time delay systems can be guaranteed.

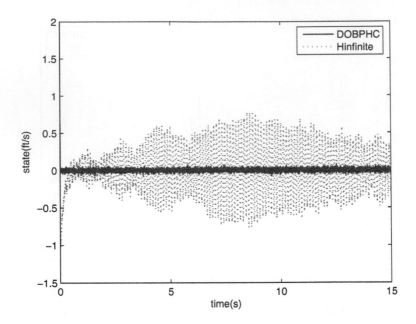

FIGURE 5.5 The comparison of system performance for the forward velocity with DOBPH_∞C and H_∞C.

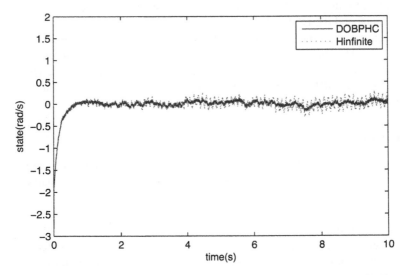

FIGURE 5.6 The comparison of system performance for angle of attack with DOBPH_∞C and H_∞C.

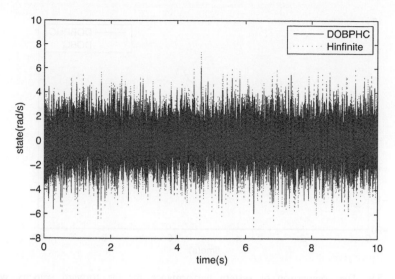

FIGURE 5.7 The comparison of system performance for the pitching velocity with DOBPH_∞C and H_∞C.

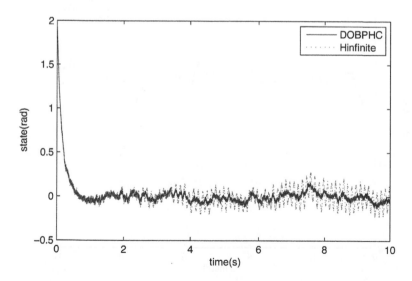

FIGURE 5.8 The comparison of system performance with DOBPH_∞C and H_∞C.

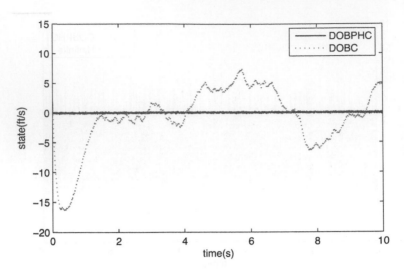

FIGURE 5.9 The comparison of system performance for the forward velocity with DOBPH_∞C and DOBC.

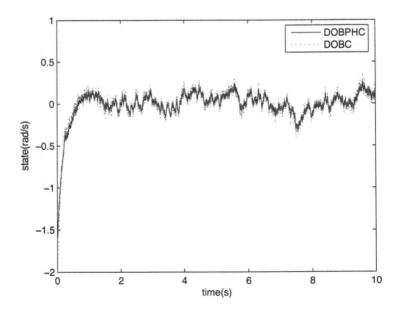

FIGURE 5.10 The comparison of system performance for the angle of attack with DOBPH_∞C and DOBC.

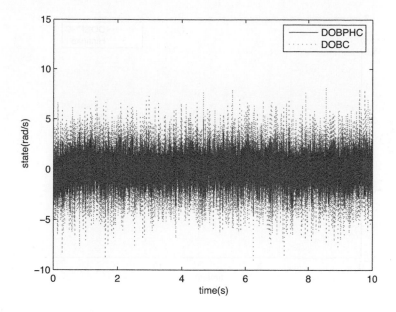

FIGURE 5.11   The comparison of system performance for the pitching velocity with DOBPH_∞C and DOBC.

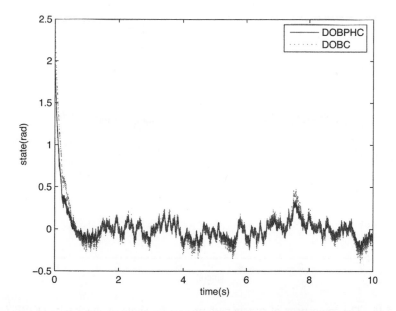

FIGURE 5.12   The comparison of system performance for the pitching angle with DOBPH_∞C and DOBC.

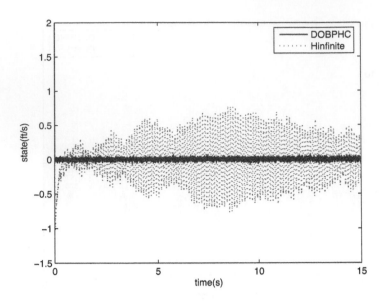

FIGURE 5.13 The comparison of system performance for the forward velocity with DOBPH_∞C and H_∞C.

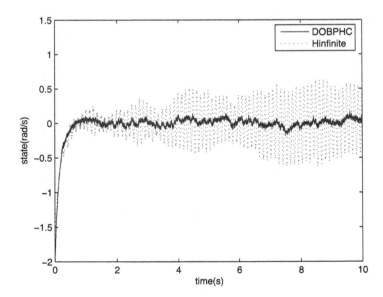

FIGURE 5.14 The comparison of system performance for angle of attack with DOBPH_∞C and H_∞C.

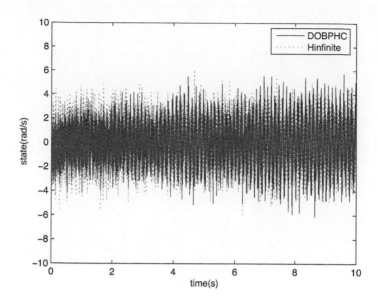

FIGURE 5.15   The comparison of system performance for the pitching velocity with DOBPH_∞C and H_∞C.

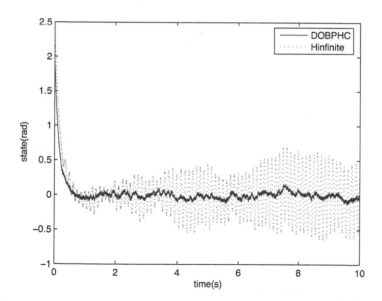

FIGURE 5.16   The comparison of system performance for the pitching angle with DOBPH_∞C and H_∞C.

6 Composite DOBC Plus H_∞ Control for Markovian Jump Systems with Multiple Disturbances

6.1 INTRODUCTION

As is well known, disturbances exist in most practical controlled processes, due to friction and load variations in mechanical and electrical systems, measurement noise, environmental disturbance, errors caused by sensors and actuators, and so on. Hence, to guarantee stability and improve performance, how to attenuate, compensate and reject the disturbances, especially for the nonlinear systems, becomes a crucial problem [30, 36, 106, 113, 159, 182, 221]. When a system encounters multiple disturbances, hierarchical/composite control strategies consisting of DOBC and another control scheme, such as robust control [202], sliding mode control [203, 201], adaptive control [93] and so on, are presented to achieve anti-disturbance performance.

On another research front, Markovian jump linear systems (MJLSs) have a strong practical background, since in practice many physical systems are subjected to abrupt variations in their structures due to random failures or repairs of components, sudden environmental disturbances, changing subsystem interconnections, and abrupt variations in the operating point of nonlinear plants. MJLSs include two parts, finite discrete jump modes and continuous states, which are governed by Markov process and differential equations. MJLSs have found wide applications in such systems as manufacturing systems, power systems, economic systems, communication systems and network-based control systems. Hence, MJLSs have been drawing continued interest from control theorists for decades. Recently, many works have been published on MJLS (see, for example, [103, 184, 200, 213, 222] and the references therein). This literature covers a wide range of problems for MJLS. To just mention a few, the problem of stability analysis and stabilization is investigated in [189, 213], various control designs are presented in [103, 128, 160, 184, 209], state estimation is reported in [59, 200, 212, 223], and the fault detection problem is considered in [222, 235]. Notice that among the above-mentioned literature of MJLS, the system has been assumed to be not subject to any disturbance or subject to just one kind of disturbance. However, this is not the case in practice, where a system is always accompanied by multiple disturbances, such as norm-bounded disturbance, disturbance with some known information and so on. Thus, how to design unconventional control strategies to guarantee system stability and achieve the desired performance in case of multiple disturbances becomes a thorny problem.

In this chapter, we propose a hierarchical/composite control methodology, that is DOBC plus H_∞ control, for Markovian jump systems with nonlinearity and multiple disturbances. Nonlinearity with known and unknown functions, respectively, are considered in this chapter. The multiple disturbances include two kinds: one is supposed to be a norm-bounded vector; the other is described by an exogenous system with perturbations. With the introduction of the notion of composite DOBC plus H_∞ control and by choosing a proper stochastic Lyapunov-Krasovskii functional, disturbance observers and special controllers are solved, such that the composite system is stochastically stable and meets certain performance requirements. Finally, numerical examples are used to illustrate the efficiency of the developed results.

6.2 PROBLEM STATEMENT AND PRELIMINARIES

Fix a probability space (Ω, F, P), where Ω is the sample space, F is the σ-algebra of subsets of the sample space and P is the probability measure on F. Under this probability space, we consider the following MJLS with nonlinearity (Σ):

$$\dot{x}(t) = A(r_t)x(t) + F(r_t)f(x(t), r_t, t) + H(r_t)d_2(t)$$
$$+ G(r_t)[u(t) + d_1(t)] \tag{6.1}$$

where $x(t) \in R^n$ is the state vector, $u(t) \in R^m$ is the control input, $f(x(t), r_t, t) \in R^q$ are nonlinear vector functions, $d_1(t) \in R^m$ is supposed to satisfy bounded conditions described as Assumption 6.1, which can represent the constant and harmonic noises. $d_2(t) \in R^q$ is another disturbance which is assumed to be an arbitrary signal in $L_2[0, \infty)$. $\{r_t\}$ is a continuous-time Markov process with right continuous trajectories and taking values in a finite set $S = \{1, 2, \dots, N\}$ with transition probability matrix $\Pi = \{\pi_{ij}\}$ given by

$$\Pr\{r_{t+\Delta} = j | r_t = i\} = \begin{cases} \pi_{ij}\Delta + o(\Delta) & \text{if } j \neq i \\ 1 + \pi_{ii}\Delta + o(\Delta) & \text{if } j = i \end{cases} \tag{6.2}$$

where $\Delta > 0$, $\lim_{\Delta \to 0}(o(\Delta)/\Delta) = 0$, $\pi_{ij} \geq 0$ is the transition rate from i at time t to j at time $t + \Delta$, and $\pi_{ii} = -\sum_{j=1, j \neq i}^{N} \pi_{ij}$.

Assumption 6.1 *The disturbance $d_1(t)$ is the control input path that can be formulated by the following exogenous system*

$$\begin{cases} \dot{\omega}(t) = W(r_t)\omega(t) + M(r_t)\delta(t) \\ d_1(t) = V(r_t)\omega(t) \end{cases} \tag{6.3}$$

where $W(r_t) \in R^{r \times r}$, $M(r_t) \in R^{r \times l}$, $V(r_t) \in R^{m \times r}$ are proper known matrices. $\delta(t) \in R^l$ is the additional disturbance which results from the perturbations and uncertainties in the exogenous system. It is also supposed that $\delta(t)$ belong to $L_2[0, \infty)$. In many cases, system disturbance can be described as a dynamical system with unknown *parameters and initial conditions, and can also be used to include the unmodeling error and system perturbations [85].*

In this chapter, we make the following assumption on the nonlinear functions in the system (6.1).

Assumption 6.2

1. *The nonlinear term $f(0, r_t, t) = 0$*
2. *For the given constant weighting matrices $U(r_t)$,*

$$\|f(x_1(t), r_t, t) - f(x_2(t), r_t, t)\| \leq \|U(r_t)(x_1(r_t, t) - x_2(r_t, t))\|$$

The following assumption is a necessary condition for the DOBC formulation.

Assumption 6.3 *The Pair $(A(r_t), G(r_t))$ is controllable and $(W(r_t), G(r_t)V(r_t))$ is observable.*

For notational simplification, $A(r_t)$ is denoted by A_i, and accordingly $G(r_t)$ and $H(r_t)$ are denoted by G_i and H_i, respectively, and so on.

In this chapter, we suppose that all of the system states are available. Then, only $d_1(t)$ is required to be estimated, and thus a reduced-order observer can be used. Here, we construct the reduced-order observers for $d_1(t)$ for the case with known nonlinearity and unknown nonlinearity, respectively. We then design a disturbance observer and a special controller so that the disturbance can be rejected and attenuated simultaneously, and the stochastical stability of the resulting composite system can also be guaranteed.

6.3 COMPOSITE DOBC PLUS H_∞ CONTROL FOR THE CASE WITH KNOWN NONLINEARITY

The disturbance observer is formulated as

$$\begin{cases} \hat{d}_1(t) &= V_i \hat{\omega}(t) \\ \hat{\omega}(t) &= v(t) - L_i x(t) \\ \dot{v}(t) &= (W_i + L_i G_i V_i)(v(t) - L_i x(t)) \\ &\quad + L_i(A_i x(t) + F_i f_i(x(t), t) + G_i u(t)) \end{cases} \tag{6.4}$$

In the DOBC scheme, the control can be constructed as

$$u(t) = -\hat{d}_1(t) + K_i x(t) \tag{6.5}$$

where $\hat{d}_1(t) \in R^r$ is the estimation of $d_1(t)$, $K_i \in R^{m \times n}$ is the controller gain, and $L_i \in R^{r \times n}$ is the observer gain.

Remark 6.1 A special form of observer-based controller is constructed in (6.5). Unlike the conventional observer-based control scheme $u(t) = K\rho(t)$ with K is the state-feedback gain and $\rho(t)$ is the estimation of the original unknown state, here the control law (6.5) includes two parts: one is the negative of the estimation of the disturbance $d_1(t)$ in (6.3), and the other is the classical mode-dependent state-feedback control laws. Obviously, with the unconventional scheme (6.5), the disturbance $d_1(t)$ generated by an exogenous system (6.3) can be compensated for through the first part of the scheme, while the second plays a role in guaranteeing the dynamical system stability and meeting performance requirements.

The estimation error is denoted as

$$e_\omega(t) = \omega(t) - \hat{\omega}(t) \tag{6.6}$$

Based on (6.1), (6.3), (6.4) and (6.6), it is shown that the error dynamics satisfies

$$\dot{e}_\omega(t) = (W_i + L_i G_i V_i)e_\omega(t) + M_i\delta(t) + L_i H_i d_2(t) \tag{6.7}$$

Combining (6.1), (6.5) and (6.7), the composite system yields

$$\dot{\xi}(t) = \bar{A}_i(t)\xi(t) + \bar{F}_i f_i(\xi(t),t) + \bar{H}_i d(t) \tag{6.8}$$

with

$$\xi(t) = \begin{bmatrix} x(t)^T & e_\omega(t)^T \end{bmatrix}^T, d(t) = \begin{bmatrix} d_2(t)^T & \delta(t)^T \end{bmatrix}^T$$

$$f_i(\xi(t),t) = f_i(x(t),t)$$

and

$$\bar{A}_i = \begin{bmatrix} A_i + G_i K_i & G_i V_i \\ 0 & W_i + L_i G_i V_i \end{bmatrix},$$

$$\bar{F}_i = \begin{bmatrix} F_i \\ 0 \end{bmatrix}, \bar{H}_i = \begin{bmatrix} H_i & 0 \\ L_i H_i & M_i \end{bmatrix}$$

The reference output is set to be

$$z(t) = C_{1i}x(t) + C_{2i}e_\omega(t) = \bar{C}_i\xi(t) \tag{6.9}$$

with $\bar{C}_i = \begin{bmatrix} C_{1i} & C_{2i} \end{bmatrix}$.

Remark 6.2 According to the composite system (6.8), $d(t) \in L_2[0,\infty)$ can be deduced, due to $\delta(t) \in L_2[0,\infty)$ and $d_2(t) \in L_2[0,\infty)$. Hence, to attenuate the disturbance $d(t)$, H_∞ control scheme is a good choice for the analysis and synthesis of dynamical systems.

Now, with Remark 6.1 and Remark 6.2, we give the notion of the composite DOBC plus H_∞ control scheme.

Definition 6.1 For dynamical system (6.1) with multiple disturbances $d_1(t)$ and $d_2(t)$, $d_2(t)$ is supposed to be H_2 norm-bounded vector; $d_1(t)$ is described by an exogenous system with H_2 norm-bounded perturbations $\delta(t)$. Estimating disturbance $d_1(t)$ with disturbance observer (6.4), then based on the output of the observer, constructing a controller with the special form (6.5), we get the composite system (6.8) with H_2 norm-bounded $d(t)$. And hence, the synthesis of the problem can be deduced to H_∞ control problem for the composite system (6.8) and (6.9). We name such scheme as composite DOBC plus H_∞ control scheme.

Remark 6.3 In the composite control scheme, H_∞ control generally achieves the attenuation performance with respect to the disturbances belonging to $L_2[0,\infty)$, while DOBC is used to reject the influence of the disturbance with some known information.

Before proposing the main problem, we give the following definition based on [23].

Definition 6.2 The Markovian jump system in (6.1) under (6.3) with $\delta(t) = d_2(t) = 0$ is said to be stochastic stable if there exists a matrix $M > 0$ such that for any initial states x_0, and initial mode $r_0 \in S$,

$$\lim_{T \to \infty} \mathbf{E} \left\{ \int_0^T x^T(t,\phi,r_0)x(t,x_0,r_0)dt \mid x_0,r_0 \right\} \leq x_0^T M x_0 \tag{6.10}$$

where $x(t,x_0,r_0)$ denotes the solution of the system (6.1) with (6.3) at the time t under the initial condition x_0 and r_0, and x_0 represents $x(t,x_0,r_0)$ at $t = 0$.

Therefore, the composite DOBC plus H_∞ control problem for system (6.1) with (6.3) can be formulated as follows.

Composite DOBC plus H_∞ Control Problem: Given the Markovian jump nonlinear system (6.1) with (6.3), design a reduced-order observer of the form (6.4) and controller of the form (6.5) such that the following requirements are satisfied:

1. The composite system in (6.8) and (6.9) with $d_2(t) = 0$ is stochastically stable.
2. Under the zero initial conditions, the following inequality holds:

$$\|z(t)\|_2 < \gamma\|d(t)\|_2 \tag{6.11}$$

for all nonzero $(t) \in L_2[0,\infty)$, where $\gamma > 0$ is a prescribed scalar, and $\|z(t)\|_2 = \int_0^\infty z^T(t)z(t)dt$.

In the following, we will present a sufficient condition in terms of linear matrix inequalities (LMIs), under which the augmented system (6.8) and (6.9) is stochastically stable and satisfies the H_∞ performance inequalities (6.11).

Theorem 6.1

Consider system (6.1) with the disturbance (6.3) under Assumptions 6.2 and 6.3. Given parameters $\lambda_i > 0$ and $\gamma > 0$, there exists a disturbance observer in the form of (6.4), and exists a controller in the form of (6.5) such that the augmented system in (6.8) and (6.9) is stochastically stable and satisfies the H_∞ performance inequalities (6.11) if there exist matrices $Q_i > 0$, X_i and Y_i such that for $i = 1, 2, \ldots, N$,

$$\begin{bmatrix} \Pi_{11i} & G_iV_i & F_i & \Pi_{14i} & Q_iC_{1i}^T & Q_iU_i^T & \Xi_i \\ * & \Pi_{22i} & 0 & \Pi_{24i} & C_{2i}^T & 0 & 0 \\ * & * & -\frac{1}{\lambda_i^2}I & 0 & 0 & 0 & 0 \\ * & * & * & -\gamma^2 I & 0 & 0 & 0 \\ * & * & * & * & -I & 0 & 0 \\ * & * & * & * & * & -\lambda_i^2 I & 0 \\ * & * & * & * & * & * & \Lambda_i \end{bmatrix} < 0 \qquad (6.12)$$

with

$$\Xi_i = \begin{bmatrix} Q_i & \cdots & Q_i & \cdots & Q_i \end{bmatrix}_{N-1}$$

$$\Lambda_i = -\text{diag}\left\{ \pi_{i1}^{-1}Q_1, \ldots, \pi_{ij}^{-1}Q_j, \ldots, \pi_{iN}^{-1}Q_N \right\}_{j\neq i}$$

$$\Pi_{11i} = A_iQ_i + Q_iA_i^T + G_iX_i + X_i^TG_i^T + \pi_{ii}Q_i$$

$$\Pi_{22i} = P_{2i}W_i + Y_iG_iV_i + (P_{2i}W_i + Y_iG_iV_i)^T + \bar{P}_{2i}$$

$$\Pi_{14i} = \begin{bmatrix} H_i & 0 \end{bmatrix}, \Pi_{24i} = \begin{bmatrix} Y_iH_i & P_{2i}M_i \end{bmatrix}$$

$$\bar{P}_{2i} = \sum_{j=1}^{N} \pi_{ij}P_{2j}$$

Moreover, if the above conditions are feasible, the gains of the desired observer in the form of (6.4) and the desired controller in the form of (6.5) can be given by

$$K_i = X_iQ_i^{-1}, \quad L_i = P_{2i}^{-1}Y_i \qquad (6.13)$$

∎

Proof. Define a Lyapunov functional candidate as follows:

$$V(\xi(t), r_t, t) = V_1(\xi(t), r_t, t) + V_2(\xi(t), r_t, t) \qquad (6.14)$$

with

$$\begin{cases} V_1(\xi(t),r_t,t) &= \xi^T(t)P_i\xi(t) \\ V_2(\xi(t),r_t,t) &= \frac{1}{\lambda_i^2}\int_0^t \|U_ix(\tau)\|^2 - \|f_i(x(\tau),\tau)\|^2 d\tau \end{cases}$$

with $P_i > 0, i \in S$.
Define

$$P_i = \begin{bmatrix} P_{1i} & 0 \\ 0 & P_{2i} \end{bmatrix} \tag{6.15}$$

with $P_{1i} \geq 0$ and $P_{2i} \geq 0$. Let A be the weak infinitesimal generator of the random process $\{\xi(t),r_t\}$. Then, for each $r_t = i, i \in S$, it can be shown that

$$\begin{aligned} AV_1(\xi(t),i,t) &= \xi^T(t)(P_i\bar{A}_i + \bar{A}_i^T P_i)\xi(t) + \xi^T(t)\bar{P}_i\xi(t) \\ &\quad +2\xi^T(t)P_i\bar{F}_if_i(x(t),t) + 2\xi^T(t)P_i\bar{H}_id(t) \end{aligned} \tag{6.16}$$

$$AV_2(\xi(t),i,t) = \frac{1}{\lambda_i^2}x^T(t)U_i^T U_ix(t) - \frac{1}{\lambda_i^2}f_i^T(x(t),t)f_i(x(t),t) \tag{6.17}$$

Combining (6.15), (6.16) and (6.17), we can derive

$$\begin{aligned} AV(\xi(t),i,t) &= AV_1(\xi(t),i,t) + AV_2(\xi(t),i,t) \\ &= \eta^T(t)\begin{bmatrix} \Phi_{1i} & P_{1i}G_iV_i & P_{1i}F_i & \Phi_{2i} \\ * & \Phi_{3i} & 0 & \Phi_{4i} \\ * & * & -\frac{1}{\lambda_i^2}I & 0 \\ * & * & * & 0 \end{bmatrix}\eta(t) \end{aligned} \tag{6.18}$$

with

$$\eta(t) = \begin{bmatrix} x^T(t) & e_\omega^T(t) & f_i^T(t,x(t)) & d^T(t) \end{bmatrix}^T$$

and

$$\Phi_{1i} = P_{1i}(A_i + G_iK_i) + (A_i + G_iK_i)^T P_{1i} + \bar{P}_{1i} + \frac{1}{\lambda_i^2}U_i^T U_i$$

$$\Phi_{2i} = \begin{bmatrix} P_{1i}H_i & 0 \end{bmatrix}, \Phi_{4i} = \begin{bmatrix} P_{2i}L_iH_i & P_{2i}M_i \end{bmatrix}$$

$$\Phi_{3i} = P_{2i}(W_i + L_iG_iV_i) + (W_i + L_iG_iV_i)^T P_{2i} + \bar{P}_{2i}$$

$$\bar{P}_{1i} = \sum_{j=1}^N \pi_{ij}P_{1j}, \quad \bar{P}_{2i} = \sum_{j=1}^N \pi_{ij}P_{2j}$$

Consider the following index

$$J(T) = \mathbf{E}\left\{ \int_0^T \left[z^T(t)z(t) - \gamma^2 d^T(t)d(t) \right] dt \right\}$$

Then, under the zero initial conditions, it follows from (6.9) and (6.18) that

$$
\begin{aligned}
J(T) &= \mathbf{E}\left\{ \int_0^T \left[z^T(t)z(t) - \gamma^2 d^T(t)d(t) \right] dt \right\} + \mathbf{E}V(\xi(T), i, T) \\
&= \mathbf{E}\left\{ \int_0^T \left[z^T(t)z(t) - \gamma^2 \omega^T(t)\omega(t) + AV(\xi(t), r_t = i) \right] dt \right\} \\
&= \mathbf{E}\left\{ \int_0^T \eta^T(t)\Theta_i\eta(t)dt \right\}
\end{aligned}
\tag{6.19}
$$

with

$$
\Theta_i = \begin{bmatrix}
\Phi_{1i} + C_{1i}^T C_{1i} & P_{1i}G_iV_i + C_{1i}^T C_{2i} & P_{1i}F_i & \Phi_{2i} \\
* & \Phi_{3i} + C_{2i}^T C_{2i} & 0 & \Phi_{4i} \\
* & * & -\frac{1}{\lambda_i^2}I & 0 \\
* & * & * & -\gamma^2 I
\end{bmatrix}
$$

with $\eta(t)$, Φ_{1i}, Φ_{2i}, Φ_{3i} and Φ_{4i} is defined in (6.18).

Now, we begin to verify that if (6.12) holds, then $\Theta_i < 0$.

Using Schur complement to inequalities (6.12), we obtain the following inequalities

$$
\begin{bmatrix}
\Pi_{11i} + \sum_{j=1, j\neq i}^{N} \pi_{ij}Q_iQ_j^{-1}Q_i & G_iV_i & F_i & \Pi_{14i} & Q_iC_{1i}^T & Q_iU_i^T \\
* & \Pi_{22i} & 0 & \Pi_{24i} & C_{2i}^T & 0 \\
* & * & -\frac{1}{\lambda_i^2}I & 0 & 0 & 0 \\
* & * & * & -\gamma^2 I & 0 & 0 \\
* & * & * & * & -I & 0 \\
* & * & * & * & * & -\lambda_i^2 I
\end{bmatrix}
$$

$$< 0 \tag{6.20}$$

with Π_{11i}, Π_{14i}, Π_{22i}, and Π_{24i} are defined in (6.12).
Define

$$P_{1i} = Q_i^{-1}, X_i = K_iP_{1i}^{-1}, Y_i = P_{2i}L_i \tag{6.21}$$

then perform a congruence transformation to (6.20) by $\text{diag}\{P_{1i}, I, I, I, I, I\}$, we readily obtain the following inequalities

$$
\begin{bmatrix}
\bar{\Phi}_{1i} & P_{1i}G_iV_i & P_{1i}F_i & \Phi_{2i} & C_{1i}^T & U_i^T \\
* & \Phi_{3i} & 0 & \Phi_{4i} & C_{2i}^T & 0 \\
* & * & -\frac{1}{\lambda_i^2}I & 0 & 0 & 0 \\
* & * & * & -\gamma^2 I & 0 & 0 \\
* & * & * & 0 & -I & 0 \\
* & * & * & 0 & 0 & -\lambda_i^2 I
\end{bmatrix} < 0 \tag{6.22}
$$

with

$$\bar{\Phi}_{1i} = P_{1i}(A_i + G_i K_i) + (A_i + G_i K_i)^T P_{1i} + \bar{P}_{1i}$$

Using Schur complement to (6.22), we can readily derive $\Theta_i < 0$.

Thus, $J(T) \leq 0$ by taking (6.19) into account. Under the zero initial conditions and for any nonzero $d(t) \in L_2(0, \infty)$, letting $T \to \infty$, we obtain $\|z(t)\|_2 \leq \gamma \|d(t)\|_2$.

Moreover, based on (6.21), the gain of the observer (6.4) and the gain of the controller (6.5) is given by (6.13). The proof is completed.   □

Remark 6.4 It is worth noting that if the composite MJLS in (6.8) and (6.9) guarantees H_∞ disturbance attenuation level γ according to Theorem 6.1, then the stochastic stability of the composite system with $d(t) = 0$ is also guaranteed. This is briefly shown as follows. First, we define the Lyapunov-Krasovskii function as in (6.14). Then, by following along the lines similar to the proof of Theorem 6.1, one can see that the weak infinitesimal to $V(\xi(t), i, t)$ along the solution of (6.8) with $d(t) = 0$ is given by

$$AV(x_t, i, t) \leq \hat{\eta}^T(t) \begin{bmatrix} \Phi_{1i} & P_{1i}G_iV_i & P_{1i}F_i \\ * & \Phi_{3i} & 0 \\ * & * & -\frac{1}{\lambda_i^2}I \end{bmatrix} \hat{\eta}(t) \qquad (6.23)$$

with

$$\hat{\eta}(t) = \begin{bmatrix} x^T(t) & e_\omega^T(t) & f_i^T(x(t), t) \end{bmatrix}^T$$

Again, using the similar arguments to the proof of Theorem 6.1, one can see that (6.12) guarantees

$$\begin{bmatrix} \Phi_{1i} & P_{1i}G_iV_i & P_{1i}F_i \\ * & \Phi_{3i} & 0 \\ * & * & -\frac{1}{\lambda_i^2}I \end{bmatrix} < 0$$

Finally, following along the lines similar to [223] and Definition 6.3, we have that the composite MJLS in (6.8) with $d(t) = 0$ is stochastically stable.

6.4 COMPOSITE DOBC PLUS H_∞ CONTROL FOR THE CASE WITH UNKNOWN NONLINEARITY

In this section, we suppose Assumptions 6.1, 6.2 and 6.3 hold, but nonlinear functions $f_i(x(t), t)$ are unknown. Different from Section 6.3, $f_i(\bar{x}(t), t)$ are unavailable in observer design.

In this section, we choose the following disturbance observer

$$\begin{cases} \hat{d}_1(t) &= V_i \hat{\omega}(t) \\ \hat{\omega}(t) &= v(t) - L_i x(t) \\ \dot{v}(t) &= (W_i + L_i G_i V_i)(v(t) - L_i x(t)) \\ & \quad + L_i (A_i x(t) + G_i u(t)) \end{cases} \qquad (6.24)$$

the controller can be constructed as

$$u(t) = -\hat{d}_1(t) + K_i x(t) \tag{6.25}$$

The estimation error is denoted as

$$e_\omega(t) = \omega(t) - \hat{\omega}(t) \tag{6.26}$$

Based on (6.1), (6.3), (6.24) and (6.26), it is shown that the error dynamics satisfies

$$\dot{e}_\omega(t) = (W_i + L_i G_i V_i) e_\omega(t) + M_i \delta(t) + L_i H_i d_2(t) \tag{6.27}$$

Combining (6.1), (6.5) and (6.7), the composite system yields

$$\dot{\xi}(t) = \bar{A}_i(t)\xi(t) + \bar{F}_i f_i(\xi(t), t) + \bar{H}_i d(t) \tag{6.28}$$

with

$$\xi(t) = \begin{bmatrix} x(t)^T & e_\omega(t)^T \end{bmatrix}^T, d(t) = \begin{bmatrix} d_2(t)^T & \delta(t)^T \end{bmatrix}^T$$

$$f_i(\xi(t), t) = f_i(x(t), t)$$

and

$$\bar{A}_i = \begin{bmatrix} A_i + G_i K_i & G_i V_i \\ 0 & W_i + L_i G_i V_i \end{bmatrix},$$

$$\bar{F}_i = \begin{bmatrix} F_i \\ L_i F_i \end{bmatrix}, \bar{H}_i = \begin{bmatrix} H_i & 0 \\ L_i H_i & M_i \end{bmatrix}$$

The reference output is set to be

$$z(t) = C_{1i} x(t) + C_{2i} e_\omega(t) = \bar{C}_i \xi(t) \tag{6.29}$$

with $\bar{C}_i = \begin{bmatrix} C_{1i} & C_{2i} \end{bmatrix}$.

In the following, we will present a sufficient condition in terms of LMIs, under which the augmented system (6.28) and (6.29) is stochastically stable and satisfies the H_∞ performance inequalities (6.11).

Theorem 6.2

Consider system (6.1) with disturbance (6.3) under Assumptions 6.2 and 6.3. Given parameters $\lambda_i > 0$ and $\gamma > 0$, there exists a disturbance observer in the form of (6.4), and exists a controller in the form of (6.5) such that the augmented system in (6.28) and (6.29) is stochastically stable and satisfies the H_∞ performance inequalities (6.11) if there exist matrices $Q_i > 0$, X_i and Y_i such that for $i = 1, 2, \ldots, N$,

$$
\begin{bmatrix}
\Pi_{11i} & G_i V & F_i & \Pi_{14i} & Q_i C_{1i}^T & Q_i U_i^T & \Xi_i \\
* & \Pi_{22i} & Y_i F_i & \Pi_{24i} & C_{2i}^T & 0 & 0 \\
* & * & -\frac{1}{\lambda_i^2} I & 0 & 0 & 0 & 0 \\
* & * & * & -\gamma^2 I & 0 & 0 & 0 \\
* & * & * & * & -I & 0 & 0 \\
* & * & * & * & * & -\lambda_i^2 I & 0 \\
* & * & * & * & * & * & \Lambda_i
\end{bmatrix} < 0
\tag{6.30}
$$

with Ξ_i, Λ_i, Π_{11i}, Π_{14i}, Π_{22i} and Π_{24i} are defined in the (6.12).

Moreover, if the above conditions are feasible, the gains of the desired observer in the form of (6.24) and the desired controller in the form of (6.25) can be given by

$$
K_i = X_i Q_i^{-1}, \quad L_i = P_{2i}^{-1} Y_i
\tag{6.31}
$$

∎

Proof. Comparing the system matrices in (6.28) and (6.29) with the system matrices in (6.8) and (6.9), and following along the similar arguments in Theorem 6.1, we can readily obtain Theorem 6.2. Hence, we omit the process of the proof. □

6.5 ILLUSTRATIVE EXAMPLE

In this section, we will present an illustrative example to demonstrate the effectiveness of the proposed approaches. Consider the systems in (6.1) and (6.3), which involve two modes, and the parameters of the systems are given as follows:

Mode 1:

$$
A_1 = \begin{bmatrix} -2.2 & 1.5 \\ 0 & 1.2 \end{bmatrix}, F_1 = \begin{bmatrix} 1.1 \\ 0.1 \end{bmatrix},
$$

$$
G_1 = \begin{bmatrix} -1.5 \\ 2.0 \end{bmatrix}, H_1 = \begin{bmatrix} 1.2 \\ 1.0 \end{bmatrix}
$$

$$
C_{11} = \begin{bmatrix} 0.5 & 0.1 \end{bmatrix}, C_{12} = \begin{bmatrix} 0.6 & 0.2 \end{bmatrix}
$$

$$
W_1 = \begin{bmatrix} 0 & 2.0 \\ -2.0 & 0 \end{bmatrix}, V_1 = \begin{bmatrix} 2.0 & 0 \end{bmatrix}, M_1 = \begin{bmatrix} 0.2 \\ 0.4 \end{bmatrix}
$$

Mode 2:

$$
A_2 = \begin{bmatrix} -1.9 & 0.5 \\ 0.2 & 1.2 \end{bmatrix}, F_2 = \begin{bmatrix} 0.2 \\ 0.1 \end{bmatrix},
$$

$$G_2 = \begin{bmatrix} -1.0 \\ 1.0 \end{bmatrix}, H_2 = \begin{bmatrix} 0.6 \\ 0.4 \end{bmatrix}$$

$$C_{21} = \begin{bmatrix} 1.2 & 0.1 \end{bmatrix}, C_{22} = \begin{bmatrix} 0.5 & 1.0 \end{bmatrix}$$

$$W_1 = \begin{bmatrix} 0 & 5.0 \\ -5.0 & 0 \end{bmatrix}, V_1 = \begin{bmatrix} 5.0 & 0 \end{bmatrix}, M_1 = \begin{bmatrix} 0.2 \\ 0.1 \end{bmatrix}$$

The transition probability matrix is assumed to be

$$\Pi = \begin{bmatrix} -1 & 1 \\ 1 & -1 \end{bmatrix}.$$

Assume

$$\lambda_1 = \lambda_2 = \gamma = 1, U_1 = U_2 = \begin{bmatrix} 0 & 0 \\ 0 & 1.0 \end{bmatrix}$$

Our intention here is to design an observer-based controller in the form of (6.4), (6.5), (6.24) and (6.25), for the case with known nonlinearity and unknown nonlinearity, respectively, such that the composite system is stochastically stable and satisfies prescribed performance requirements.

Case 1: with known nonlinearity

We resort to the LMI Toolbox in MATLAB to solve the LMIs in (6.12), and the gains of the desired observer and controller are given by

$$K_1 = \begin{bmatrix} -2.3236 & -7.6187 \end{bmatrix}, \quad K_2 = \begin{bmatrix} -3.0767 & -15.1046 \end{bmatrix}$$

$$L_1 = \begin{bmatrix} -0.2036 & -0.8872 \\ -0.2302 & -1.0123 \end{bmatrix}, \quad L_2 = \begin{bmatrix} -0.1557 & -1.2274 \\ -0.1274 & -0.8885 \end{bmatrix}$$

Suppose

$$f_1(t, x(t)) = f_2(t, x(t)) = x_2(t) sin(t),$$

we can find

$$\|f_i(x(t), t)\| \le \|U_i(t, x_i(t)\|, i = 1, 2.$$

Assume

$$d_2(t) = \frac{1}{5 + 10t}.$$

Given the initial condition as

$$\xi(0) = \begin{bmatrix} 0.8 & -0.5 & 0.2 & -0.1 \end{bmatrix}^T$$

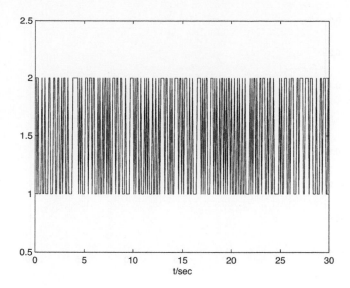

FIGURE 6.1   Switching signal.

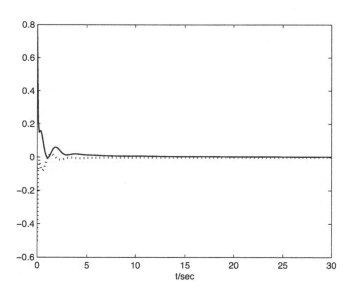

FIGURE 6.2   States of the system (6.1) for known nonlinearity.

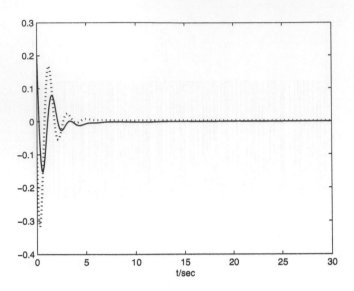

FIGURE 6.3   Estimation error (6.7) for known nonlinearity.

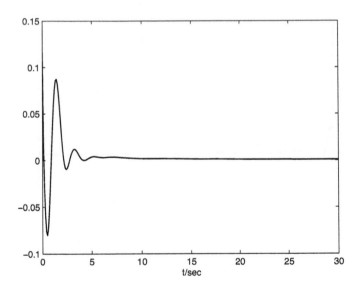

FIGURE 6.4   Output of the system (6.9) for known nonlinearity.

Figure 6.1 plots a switching signal, where '1' and '2' represent respectively the first and the second subsystem. Figure 6.2 describes the states of the system (6.1). Figure 6.3 shows the estimation error (6.7), while Figure 6.3 displays the estimation error (6.7). Figure 6.4 depicts the output (6.9) of the composite system (6.8). The simulation results clearly indicate that the composite system is stochastically stable and able to reject and attenuate multiple disturbances under the observer-based controller obtained above.

Case 2: with unknown nonlinearity

By solving the LMIs in (6.30), and the gains of the desired observer and controller are given by

$$K_1 = \begin{bmatrix} -1.6483 & -4.6944 \end{bmatrix}, \quad K_2 = \begin{bmatrix} -2.1860 & -9.4961 \end{bmatrix}$$

$$L_1 = \begin{bmatrix} -0.1254 & -0.6845 \\ -0.0182 & -0.1601 \end{bmatrix}, \quad L_2 = \begin{bmatrix} -0.0670 & -0.9037 \\ -0.1008 & -0.9291 \end{bmatrix}$$

Suppose $f_1(t, x(t)) = f_2(t, x(t)) = \alpha x_2(t)$ with α is a random number belonging to [0, 1], we can find

$$\|f_i(x(t), t)\| \le \|U_i(t, x_i(t))\|, i = 1, 2.$$

Given the initial condition $\xi(0)$ and $d_2(t)$ the same as *Case 1*. Figure 6.5 describes the states of the system (6.1), while Figure 6.6 shows the estimation error (6.27). Figure 6.7 depicts the output (6.29) of the composite system (6.28). The simulation results also indicate that the composite system is stochastically stable and able to reject and attenuate multiple disturbances under the observer-based controller obtained above.

6.6 CONCLUSION

In this work, composite control problems have been investigated for a class of nonlinear systems with jump parameters and multiple disturbances. The Lyapunov stability approach and the LMI technique have been applied to the analysis and design of the disturbance observer and controller for the system concerned. The designed observer and controller ensure a prescribed performance level of the resulting composite system.

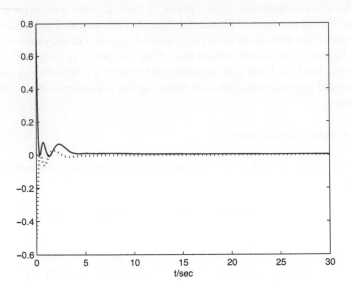

FIGURE 6.5   States of the system (6.1) for unknown nonlinearity.

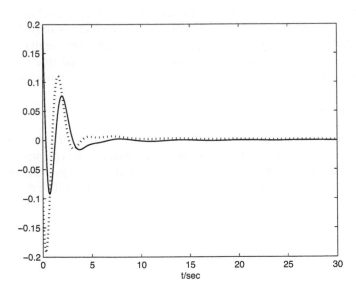

FIGURE 6.6   Estimation error (6.27) for unknown nonlinearity.

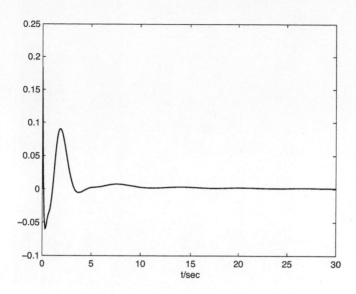

FIGURE 6.7   Output of the system (6.29) for unknown nonlinearity.

7 Robust Consensus of Multi-Agent Systems with Time Delays and Disturbances

7.1 INTRODUCTION

Consensus means that a team of agents reach an agreement on a common value by negotiating with their neighbors. Some results of consensus algorithms can be applied in many areas including cooperative control of mobile robots, unmanned air vehicles (UAVs), autonomous underwater vehicles (AUVs), automated highway systems, and so on[62, 114, 152, 163, 164, 171, 172, 173, 175].

Consensus algorithms have been extensively studied by many researchers. The study of the alignment problem involving reaching an agreement without computing any objective functions appeared in the work [114]. The theoretical framework for posing and solving consensus problems for networked dynamical systems was introduced in [164]. Consensus in the multi-agent systems was presented in [152, 173] with directed information flow in networks. [171, 172] proposed a set of second-order consensus algorithms, which include fundamental consensus algorithm and consensus algorithm with information feedback under directed networks. [100, 101] proposed an observer to estimate the variable leader's velocity. [50, 220] proposed an optimal decentralized/centralized estimator to implement the consensus. The studies for the consensus algorithms of multi-agent systems were shown in [130, 174, 176, 211] and the references therein.

Due to the finite speeds of transmission and spreading, as well as traffic congestion, there are usually time delays in spreading and communication in reality. Therefore, it is very important to study the delay effect on convergence of consensus protocols. In [164], a delay-dependent condition is proposed for uniformly constant delay in undirected networks, which depends on the second largest eigenvalue of graph Laplacian matrix. Uniformly constant delay was considered in directed networks in [151], where the stability result of the system was robust with respect to an arbitrary delay. In [11, 102, 129, 191, 217], multi-agent systems with the heterogeneous communication delays and time-varying delays were studied for undirected networks and directed networks, respectively.

The main focus of this chapter is on team cooperation or consensus for multi-agent systems with time delays and exogenous disturbances. [153] and [155] pioneered the development of disturbance observers (DOs) for robot control independently. [162] improved a linear disturbance observer in robots using the information of nonlinear inertial coupling dynamics. [32, 58] developed a nonlinear disturbance observer based control (NLDOBC) for unknown constant and applied it to a two-link manipulator. After that, DOs have been applied in many mechanistic systems including disk drives, machining centers, dc/ac motors, manipulators, etc. [85, 201].

In this chapter, we apply the method of DOs and pinning control to stabilize the states of multi-agent systems with time delays and exogenous disturbances. To our knowledge, this is the first time the method of DOs and pinning control has been applied to study consensus in delayed multi-agent systems with disturbances.

7.2 PRELIMINARIES

A weighted connected graph $G = \{V, E, A\}$ of order n consists of a set of vertices $V = \{v_1, v_2, ..., v_n\}$, a set of edges $E \subseteq V \times V$ and an adjacency matrix $A = [a_{ij}] \in R^{n \times n}$ with weighted adjacency elements $a_{ij} \geq 0$. The node indexes belong to a finite index set $I = \{1, 2, ..., n\}$. An edge of the weighted diagraph G is denoted by $e_{ij} = (v_i, v_j) \in E$ indicating that node j can influence node i, but not vice versa. We assume that the adjacency element $a_{ij} > 0$ when $e_{ij} \in E$, otherwise, $a_{ij} = 0$. The set of neighbors of node i is denoted by $N_i = \{j \in I : a_{ij} > 0\}$.

Let G be a weighted graph without self-loops, i.e., $a_{ii} = 0$, and matrix D be a diagonal matrix with the elements D_i along the diagonal, where $D_i = \sum_{j=1}^{n} a_{ij}$ is the out-degree of the node i. The Laplacian matrix of the weighted graph is defined as $L = D - A$.

For any node i and j, if there are a set of nodes $\{k_1, k_2, ...k_l\}$ satisfying $a_{ik_1} > 0$, $a_{k_1 k_2} > 0$, ..., $a_{k_l j} > 0$, node j is said to be reachable from i. If node i is reachable from every other node in the graph, it is said to be globally reachable. If each node is globally reachable in the directed graph, the graph is said to be strongly connected.

[164] showed that the following linear dynamic system solves a consensus problem,

$$\dot{x}_i(t) = - \sum_{j \in N_i} a_{ij}(x_i(t) - x_j(t)) \tag{7.1}$$

where x_i is the state of agent i. For the interconnected graph, the algorithm (1) can reach the average consensus, i.e., consensus state

$$x^* = \frac{1}{n} \sum_{i=1}^{n} x_i(0).$$

Suppose that agent i receives a message sent by its neighbor j after a time delay of τ. This is equivalent to a network with a uniform one-hop communication time delay. The following consensus algorithm

$$\dot{x}_i(t) = - \sum_{j \in N_i} a_{ij}(x_i(t - \tau) - x_j(t - \tau)). \tag{7.2}$$

was proposed in [164] to reach an average-consensus for undirected graph G, if

$$\tau < \tau_{max} = \frac{\pi}{2\lambda_n},$$

where λ_n is the largest eigenvalue of Laplacian L.

7.3 CONSENSUS ALGORITHM OF MULTI-AGENT SYSTEMS WITH EXOGENOUS DISTURBANCES

7.3.1 MULTI-AGENT SYSTEMS WITH DISTURBANCES

The consensus state of system (7.1) is affected by the initial values of multi-agent systems. If the initial values are set at random, the consensus state cannot be converged to an expected moving track. Moreover, unmodeled dynamics and parametric variations as well as external disturbances widely exist in practical processes, which will influence the consensus of a system. In this chapter, we consider a delayed system (7.2) with exogenous disturbances, and develop a controller to bring the consensus state of multiple agents to an expected moving track. The multi-agent systems with time delays and disturbances are described by, for all $i \in I$,

$$
\begin{cases}
\dot{x}_i(t) &= -\sum_{j=1}^n a_{ij}(x_i(t-\tau) - x_j(t-\tau)) + u_i(t) \\
&\quad + g(x_i)(d_i(t) + \omega_i(t)), \\
y_i(t) &= x_i(t),
\end{cases}
\tag{7.3}
$$

where $x_i \in R^{m1}$, $u_i \in R^{m1}$ and $d_i \in R^{m1}$ are the state, control input and external disturbance, respectively, $g(.)$ is a smooth valued function, ω_i is the disturbance uncertainties in the systems. In order to calculate easily, we suppose $m1 = 1$. It is supposed that the disturbance d_i, for $i = 1, ..., n$, is generated by a linear exogenous system

$$
\begin{cases}
\dot{\xi}_i(t) &= A\xi_i(t), \\
d_i(t) &= C\xi_i(t),
\end{cases}
\tag{7.4}
$$

where $\xi_i \in R^m$ is the internal state of the exogenous system, A and C are matrices with appropriate dimensions and (A, V) is observable.

Suppose that we want to stabilize multi-agent systems (7.3) onto an expected consensus state defined by

$$
x_1 = ... = x_n = \bar{x}.
\tag{7.5}
$$

where $\bar{x}$ is an expected consensus state for multiple agents. In this chapter, $\bar{x}$ is supposed as a constant invariable state.

In this chapter, we discuss the consensus of delayed multi-agent dynamical systems with exogenous disturbance. The objective of this chapter is to design DOBC techniques such that the expected consensus (7.5) on the composite controller is achieved asymptotically. The block diagram of systems (7.3) is shown in Figure 7.1. As seen from this diagram, the composite controller consists of two parts: a controller with poor disturbance attenuation ability and a disturbance observer.

The procedure of designing DOBC consists of two stages. In the first stage, the controller is designed under the assumption that there is no disturbance or the disturbance is measurable. In the second stage, a disturbance observer is designed to estimate the disturbance.

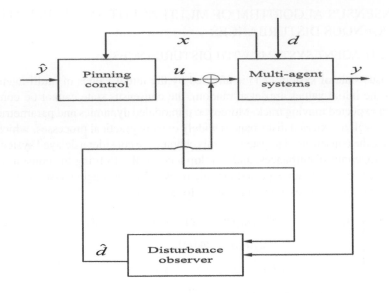

FIGURE 7.1   Structure of complex dynamical networks with disturbance observer.

7.3.2 PINNING CONTROL FOR MULTI-AGENT SYSTEMS WITHOUT DIS-TURBANCES

Supposing there is no disturbance or the disturbance is measurable, we discuss the consensus of system (7.3) by designing a controller. To realize the expected track of system (7.3), some controllers may be designed and applied to force the system to reach consensus. The simplest way to design may be to add controllers to all nodes, that is, every node is driven by the expected states directly. However, it is literally impossible to append controllers to all agents because of economic expense and operational feasibility in a large-scale network. To reduce the number of controlled nodes, some local feedback injections may be applied to a fraction of network nodes, which is known as "pinning control." Feedback pinning control has been a common technique for the control of spatiotemporal chaos in regular dynamical networks [135, 228]. Here, the pinning strategy is applied on a small fraction $\delta(0 < \delta \ll 1)$ of the nodes. Suppose that the nodes $i_1, i_2, ..., i_l$ are selected, where $l = \lfloor \delta N \rfloor$ stands for the smaller but nearest integer to the real number δN. Without loss of generality, rearrange the order of the nodes in the network, and let the first l nodes be controlled. Thus, the pinning controlled system (7.3) without disturbance can be described by

$$\begin{cases} \dot{x}_i(t) & = -\sum_{j=1}^{n} a_{ij}(x_i(t-\tau) - x_j(t-\tau)) + u_i(t), i = 1, ..., l, \\ \dot{x}_i(t) & = -\sum_{j=1}^{n} a_{ij}(x_i(t-\tau) - x_j(t-\tau)), i = l+1, ..., n, \end{cases} \tag{7.6}$$

with pinning control

$$u_i(t) = -b_i(x_i(t-\tau) - \bar{x}),\qquad(7.7)$$

where the control gain $b_i > 0$, for all $i = 1, ..., l$. Let $B = \text{diag}\{b_1, ..., b_l, b_{l+1}, ..., b_n\}$ where $b_i = 0$ for all $i = l+1, ..., n$.

Lemma 7.1

Assume the directed interconnection graph of n agents is strongly connected. Then, the matrix $L + B$ is positive definite, where L is the Laplacian matrix of the interconnection graph of n agents [174]. ∎

Lemma 7.2

(Schur complement.) Let M, P, and Q be given matrices such that $Q > 0$

$$\begin{bmatrix} P & M \\ M^T & -Q \end{bmatrix} < 0 \iff P + MQ^{-1}M^T < 0.$$

∎

Theorem 7.1

Assume multi-agent systems are composed of n agents with a strongly connected directed graph. Then, based on pinning control (7.7), the multi-agent systems (7.6) without disturbance can asymptotically reach the expected moving consensus for $\tau \in (0, \tau^*)$, if there exist matrices $Q > 0$, $R > 0$, $\tau^* > 0$, satisfying

$$\begin{bmatrix} -2\bar{L}+Q & \bar{L}^T & 0 & 0 \\ \bar{L} & -\frac{R}{\tau^*} & 0 & 0 \\ 0 & 0 & -Q & \bar{L}^T \\ 0 & 0 & \bar{L} & -\frac{R^{-1}}{\tau^*} \end{bmatrix} < 0.\qquad(7.8)$$

∎

Proof. Let $\delta x_i = x_i - \bar{x}$, and $\delta x = [\delta x_1, ..., \delta x_n]^T$. Then, the multi-agent systems (7.6) with pinning control (7.7) can be rewritten as

$$\delta \dot{x}(t) = -\bar{L}\delta x(t-\tau),\qquad(7.9)$$

where $\bar{L} = L + B$. Let Lyapunov function

$$V_1(t) = \delta x^T(t)\delta x(t) + \int_{t-\tau}^{t} \delta x^T(s)Q\delta x(s)ds + \int_{-\tau}^{0}\int_{t+\theta}^{t} \delta \dot{x}^T(s)R\delta \dot{x}(s)dsd\theta$$

The derivative of $V_1(t)$ along the trajectories of (7.9) gives

$$
\begin{aligned}
\dot{V}_1(t) = \ & -2\delta x^T(t)\bar{L}\delta x(t-\tau) + \delta x^T(t)Q\delta x(t) \\
& -\delta x^T(t-\tau)Q\delta x(t-\tau) + \tau\delta\dot{x}^T(t)R\delta\dot{x}(t) \\
& -\int_{t-\tau}^{t}\delta\dot{x}^T(s)R\delta\dot{x}(s)ds \\
\leq \ & -2\delta x^T(t)\bar{L}\delta x(t) + \tau\delta x^T(t)\bar{L}^T R^{-1}\bar{L}\delta x(t) \\
& +\delta x^T(t)Q\delta x(t) - \delta x^T(t-\tau)Q\delta x(t-\tau) \\
& +\tau\delta x^T(t-\tau)\bar{L}^T R\bar{L}\delta x(t-\tau) \\
\leq \ & -2\delta x^T(t)\bar{L}\delta x(t) + \tau^*\delta x^T(t)\bar{L}^T R^{-1}\bar{L}\delta x(t) \\
& +\delta x^T(t)Q\delta x(t) - \delta x^T(t-\tau)Q\delta x(t-\tau) \\
& +\tau^*\delta x^T(t-\tau)\bar{L}^T R\bar{L}\delta x(t-\tau) \\
= \ & \begin{pmatrix} \delta x(t) \\ \delta x(t-\tau) \end{pmatrix}^T \Sigma \begin{pmatrix} \delta x(t) \\ \delta x(t-\tau) \end{pmatrix}.
\end{aligned}
\tag{7.10}
$$

where

$$
\Sigma = \begin{bmatrix} -2\bar{L}+Q+\tau^*\bar{L}^T R^{-1}\bar{L} & 0 \\ 0 & -Q+\tau^*\bar{L}^T R\bar{L} \end{bmatrix}
$$

Following Lemma 7.2 and the condition (7.8), we can obtain $\Sigma < 0$. Then, the system (7.9) is asymptotically stable by applying Lyapunov stability theorem. The consensus of multi-agent system (7.6) without disturbances can reach the expected consensus state. The proof is finished. $\qquad\square$

7.3.3 DISTURBANCE OBSERVERS FOR MULTI-AGENT SYSTEMS WITH DISTURBANCES

It is supposed that the disturbance is generated by a linear exogenous system (7.4). A nonlinear disturbance observer is proposed to estimate the unknown disturbance d_i in system (7.3), for all $i \in I$,

$$
\begin{cases}
\dot{z}_i = & Az_i - g(x_i)KCz_i + AKx_i - Kg(x_i)CKx_i \\
& +K(\sum_{j=1}^{n} a_{ij}(x_i(t-\tau) - x_j(t-\tau)) - u_i), \\
\hat{\xi}_i = & z_i + Kx_i, \\
\hat{d}_i = & C\hat{\xi}_i,
\end{cases}
\tag{7.11}
$$

where $z_i \in R^m$ is the internal state variables of the observer and $K \in R^{m\times 1}$ is a gain matrix to be designed, $\hat{\xi}_i$ and $\hat{d}_i$ are the estimates of ξ_i and d_i, respectively. Let the estimation error be defined as

$$
e_i(t) = \xi_i(t) - \hat{\xi}_i(t).
\tag{7.12}
$$

Based on Equations (7.11), (7.12), and (7.3), we can obtain

$$\dot{e}_i(t) = (A - g(x_i)KC)e_i(t) - Kg(x_i)\omega_i(t). \tag{7.13}$$

Lemma 7.3

Consider system (7.3) under the disturbance generated by exogenous system (7.4) without uncertainties. The disturbance observers (7.11) can exponentially track the disturbance if the gain matrix K is chosen such that

$$\dot{e}(t) = ((A - g(x)KC) \otimes I_n)e(t), \tag{7.14}$$

is globally exponentially stable, where $e = [e_1, e_2, ..., e_n]^T$, $I_n \in R^{n \times n}$ is identity matrix, and $\otimes$ denotes the Kronecker product. ∎

Theorem 7.2

Consider multi-agent system (7.3) with the exogenous disturbances (7.4) and disturbance uncertainties $\omega_i(t) = 0$. Assume multi-agent systems are composed of n agents with a strongly connected directed graph. Closed-loop dynamic networks under the disturbance observers (7.11) can reach expected consensus asymptotically for $\tau \in (0, \tau^*)$, if there exist appropriate positive definite matrices Q, R, P and constant number $\tau^* > 0$, such that

$$\begin{bmatrix} \Theta_{11} & \Theta_{12} \\ \Theta_{12}^T & \Theta_{22} \end{bmatrix} < 0, \tag{7.15}$$

where

$$\Theta_{11} = \begin{bmatrix} -2\bar{L} + Q + \tau^* \bar{L}^T R^{-1} \bar{L} & 0 \\ 0 & -Q + \tau^* \bar{L}^T R \bar{L} \end{bmatrix}$$

$$\Theta_{12} = [G^T, -\tau^* G^T R^T \bar{L}^T]^T, \Theta_{22} = \tau^* G^T RG + \Omega,$$

$$G = g(x)C \otimes I_n, \Omega = (A^T P + PA - 2g(x)C^T C) \otimes I_n.$$

∎

Proof. Consider multi-agent system (7.3) without uncertainties. In order to asymptotically stabilize the system for any disturbance, a part of the control effort, $u(x, d)$, shall depend on the disturbance d. Here, we apply $\hat{d}$ instead of d, let the composite controller

$$u_i(t) = -b_i((x_i(t - \tau) - \bar{x})) - g(x_i)\hat{d}_i(t), \tag{7.16}$$

where $\bar{x}$ is an expected consensus state for multiple agents. Substituting (7.16) into system (7.3) obtains

$$
\begin{aligned}
\dot{x}_i(t) \;=\; & -\sum_{j=1}^{n} a_{ij}((x_i(t-\tau)-x_j(t-\tau))) \\
& -b_i((x_i(t-\tau)-\bar{x}))-g(x_i)\hat{d}_i(t)+g(x_i)d_i(t), \\
& i=1,...,n,
\end{aligned}
\tag{7.17}
$$

Let $\delta x_i = x_i - \bar{x}$, and $\delta x = [\delta x_1,...,\delta x_n]^T$. We have

$$
\delta \dot{x}(t) = -\bar{L}\delta x(t-\tau)+Ge(t).
\tag{7.18}
$$

Now, we discuss the closed-loop system including the multi-agent system (7.18) and the observer error dynamics (7.13) with disturbance uncertainties $\omega_i(t)=0$,

$$
\dot{e}_i(t) = \bar{A}e_i(t),
\tag{7.19}
$$

where $\bar{A} = (A - g(x)KC)$. Let a Lyapunov function

$$
V(t) = V_1(t) + \sum_{i=1}^{n} e_i^T(t)Pe_i(t)
$$

where matrix $P > 0$. The derivative of V(t) along the trajectories of (7.18-7.19) gives

$$
\begin{aligned}
\dot{V}(t) \;=\; & -2\delta x^T(t)\bar{L}\delta x(t-\tau)+\delta x^T(t)Q\delta x(t) \\
& -\delta x^T(t-\tau)Q\delta x(t-\tau)+\tau\delta \dot{x}^T(t)R\delta \dot{x}(t) \\
& -\int_{t-\tau}^{t}\delta \dot{x}^T(s)R\delta \dot{x}(s)ds+2\delta x^T(t)Ge(t) \\
& +\sum_{i=1}^{n} e_i^T(t)P\dot{e}_i(t)+\sum_{i=1}^{n}\dot{e}_i^T(t)Pe_i(t) \\
\leq\; & -2\delta x^T(t)\bar{L}\delta x(t-\tau)+\delta x^T(t)Q\delta x(t) \\
& -\delta x^T(t-\tau)Q\delta x(t-\tau)+\tau^*\delta \dot{x}^T(t)R\delta \dot{x}(t) \\
& -\int_{t-\tau}^{t}\delta \dot{x}^T(s)R\delta \dot{x}(s)ds+2\delta x^T(t)Ge(t) \\
& +\sum_{i=1}^{n} e_i^T(t)P\dot{e}_i(t)+\sum_{i=1}^{n}\dot{e}_i^T(t)Pe_i(t) \\
\leq\; & -2\delta x^T(t)\bar{L}\delta x(t)+\tau^*\delta x^T(t)\bar{L}^T R^{-1}\bar{L}\delta x(t) \\
& +\delta x^T(t)Q\delta x(t)-\delta x^T(t-\tau)Q\delta x(t-\tau) \\
& +\tau^*\delta x^T(t-\tau)\bar{L}^T R\bar{L}\delta x(t-\tau)+\tau^* e^T(t)G^T RGe(t) \\
& -2\tau^* e^T(t)G^T R\bar{L}^T \delta x(t-\tau)+2\delta x^T(t)Ge(t) \\
& +\sum_{i=1}^{n} e_i^T(t)\bar{A}^T Pe_i(t)+\sum_{i=1}^{n} e_i^T(t)P\bar{A}e_i(t).
\end{aligned}
\tag{7.20}
$$

Applying Equation (7.13) and substituting into (7.20), we have

$$
\dot{V}(t) \leq \begin{pmatrix} \delta x(t) \\ \delta x(t-\tau) \\ e(t) \end{pmatrix}^T \Psi \begin{pmatrix} \delta x(t) \\ \delta x(t-\tau) \\ e(t) \end{pmatrix}.
\tag{7.21}
$$

where

$$\Psi = \begin{bmatrix} \Theta_{11} & \Theta_{12} \\ \Theta_{12}^T & \bar{\Theta}_{22} \end{bmatrix}, \bar{\Theta}_{22} = \tau^* G^T RG + \bar{\Omega},$$

$$\bar{\Omega} = (PA + A^T P - g(x)PKC - g(x)C^T K^T P) \otimes I_n.$$

If the gain K is defined by $K = P^{-1}C^T$, we can obtain $\Psi < 0$ from the condition (7.15). Then, the system (7.3) is asymptotically stable by applying Lyapunov stability theorem in (7.21), i.e., the consensus of the multi-agent systems can be achieved under the condition (7.15). Based on Lemma 7.2 (Schur complement), the result of Lemma 7.3 can be obtained from $\Theta_{22} < 0$, too. $\qquad\square$

Theorem 7.3

Consider multi-agent system (7.3) with the exogenous disturbances (7.4) and disturbance uncertainties $\omega_i(t) \neq 0$. Assume multi-agent systems are composed of n agents with a strongly connected directed graph. If there exist appropriate positive definite matrices Q, R, P, constant number $\tau^* > 0$ and given parameters $\gamma > 0$, such that

$$\Phi = \begin{bmatrix} \Phi_{11} & \Phi_{12} & \Phi_{13} \\ \Phi_{12}^T & \Phi_{22} & \Phi_{23} \\ \Phi_{13}^T & \Phi_{23}^T & \Phi_{33} \end{bmatrix} < 0, \tag{7.22}$$

where

$$\Phi_{11} = \begin{bmatrix} \bar{\Phi}_{11} & 0 \\ 0 & -Q + \tau^* \bar{L}^T R \bar{L} \end{bmatrix},$$

$$\bar{\Phi}_{11} = -2\bar{L} + Q + \tau^* \bar{L}^T R^{-1} \bar{L} + I_n$$

$$\Phi_{12} = [G^T, -\tau^* G^T R^T \bar{L}^T]^T, \Phi_{22} = \tau^* G^T RG + \Omega,$$

$$\Phi_{13} = [H^T, -\tau^* H^T R^T \bar{L}^T]^T,$$

$$\Phi_{33} = \tau^* H^T RH - \gamma^2 I_n, \Phi_{23} = -G,$$

$$H = g(x) \otimes I_n, G = g(x)C \otimes I_n,$$

$$\Omega = (A^T P + PA - 2g(x)C^T C) \otimes I_n.$$

Then, the closed-loop dynamical networks under the disturbance observers (7.11) can reach expected consensus asymptotically for $\tau \in (0, \tau^*)$ and satisfies

$$||y(t)||_2 \leq \gamma ||\omega(t)||_2,$$

where

$$y(t) = (y_1(t), ..., y_n(t))^T, y_i(t) = \delta x_i(t), \omega(t) = (\omega_1(t), ..., \omega_n(t))^T.$$

■

Proof. In the Throrem 7.2, a disturbance observer based control is presented by an observer to estimate the exogenous disturbance and a pinning controller such that multi-agent system (7.3) is stable in the absence of the disturbance uncertainties $\omega_i(t)$. Then, we will focus on the condition of uncertainties disturbance attenuation. The closed-loop system with uncertainties $\omega_i(t) \neq 0$ including the multi-agent system and the observer error dynamics is given by

$$
\begin{aligned}
\dot{x}_i(t) &= -\sum_{j=1}^{n} a_{ij}((x_i(t-\tau) - x_j(t-\tau))) \\
&\quad -b_i((x_i(t-\tau) - \bar{x})) - g(x_i)\hat{d}_i(t) \\
&\quad +g(x_i)d_i(t) + g(x_i)\omega_i(t), \\
\dot{e}_i(t) &= \bar{A}e_i(t) - Kg(x_i)\omega_i(t),
\end{aligned}
\tag{7.23}
$$

where

$$\bar{A} = (A - g(x)KC)$$

Let $\delta x_i = x_i - \bar{x}$, and $\delta x = [\delta x_1, ..., \delta x_n]^T$, applying Lyapunov function in the proof of Theorem 7.2,

$$
\begin{aligned}
V(t) &= \delta x^T(t)\delta x(t) + \int_{t-\tau}^{t} \delta x^T(s)Q\delta x(s)ds \\
&\quad + \int_{-\tau}^{0} \int_{t+\theta}^{t} \delta \dot{x}^T(s)R\delta \dot{x}(s)dsd\theta + \sum_{i=1}^{n} e_i^T(t)Pe_i(t)
\end{aligned}
$$

where matrix $Q > 0, R > 0, P > 0$.

The following auxiliary function is considered

$$J(x(t)) = V(t) + \int_{0}^{t} (||y(\tau)||^2 - \gamma^2 ||\omega(\tau)||^2)d\tau, \tag{7.24}$$

which satisfies $J(x(t)) = \int_{0}^{t} S(\tau)d\tau$ with the zero initial condition, where $V(t)$ is Lyapunov function,

$$y(t) = (\delta x_1(t), ..., \delta x_n(t))^T, \omega(t) = (\omega_1(t), ..., \omega_n(t))^T,$$

$$S(t) = ||y(t)||^2 - \gamma^2 ||\omega(t)||^2 + \dot{V}(t),$$

then it can be verified that

$$
\begin{aligned}
S(t) &= \sum_{i=1}^{n} y_i^T(t) y_i(t) - \gamma^2 \sum_{i=1}^{n} \omega_i^T(t) \omega_i(t) \\
&\quad -2\delta x^T(t)\bar{L}\delta x(t-\tau) + \delta x^T(t)Q\delta x(t) \\
&\quad -\delta x^T(t-\tau)Q\delta x(t-\tau) \\
&\quad +\tau\delta\dot{x}^T(t)R\delta\dot{x}(t) - \int_{t-\tau}^{t} \delta\dot{x}^T(s)R\delta\dot{x}(s)ds \\
&\quad +2\delta x^T(t)Ge(t) + 2\delta x^T(t)H\omega(t) \\
&\quad +\sum_{i=1}^{n} e_i^T(t)P\dot{e}_i(t) + \sum_{i=1}^{n} \dot{e}_i^T(t)Pe_i(t) \\
&\leq \sum_{i=1}^{n} y_i^T(t) y_i(t) - \gamma^2 \sum_{i=1}^{n} \omega_i^T(t) \omega_i(t) \\
&\quad -2\delta x^T(t)\bar{L}\delta x(t-\tau) + \delta x^T(t)Q\delta x(t) \\
&\quad -\delta x^T(t-\tau)Q\delta x(t-\tau) \\
&\quad +\tau^*\delta\dot{x}^T(t)R\delta\dot{x}(t) - \int_{t-\tau}^{t} \delta\dot{x}^T(s)R\delta\dot{x}(s)ds \\
&\quad +2\delta x^T(t)Ge(t) + 2\delta x^T(t)H\omega(t) \\
&\quad +\sum_{i=1}^{n} e_i^T(t)P\dot{e}_i(t) + \sum_{i=1}^{n} \dot{e}_i^T(t)Pe_i(t) \\
&\leq \sum_{i=1}^{n} \delta x_i^T(t)\delta x_i(t) - \gamma^2 \sum_{i=1}^{n} \omega_i^T(t)\omega_i(t) \\
&\quad -2\delta x^T(t)\bar{L}\delta x(t) + \tau^*\delta x^T(t)\bar{L}^T R^{-1}\bar{L}\delta x(t) \\
&\quad +\delta x^T(t)Q\delta x(t) - \delta x^T(t-\tau)Q\delta x(t-\tau) \\
&\quad +\tau^*\delta x^T(t-\tau)\bar{L}^T R\bar{L}\delta x(t-\tau) + \tau^* e^T(t)G^T RGe(t) \\
&\quad -2\tau^* e^T(t)G^T R\bar{L}^T\delta x(t-\tau) + 2\delta x^T(t)Ge(t) \\
&\quad +\tau^*\omega^T(t)H^T RH\omega(t) + 2\delta x^T(t)H\omega(t) \\
&\quad -2\tau^*\omega^T(t)H^T R\bar{L}^T\delta x(t-\tau) \\
&\quad +\sum_{i=1}^{n} e_i^T(t)\bar{A}^T Pe_i(t) + \sum_{i=1}^{n} e_i^T(t)P\bar{A}e_i(t) \\
&\quad -\sum_{i=1}^{n} \omega_i^T(t)gK^T Pe_i(t) - \sum_{i=1}^{n} e_i^T(t)PKg\omega_i(t) \\
&\leq \begin{bmatrix} \delta x(t) \\ \delta x(t-\tau) \\ e(t) \\ \omega(t) \end{bmatrix}^T \bar{\Phi} \begin{bmatrix} \delta x(t) \\ \delta x(t-\tau) \\ e(t) \\ \omega(t) \end{bmatrix}.
\end{aligned}
$$

where

$$
\bar{\Phi} = \begin{bmatrix} \Phi_{11} & \Phi_{12} & \Phi_{13} \\ \Phi_{12}^T & \bar{\Phi}_{22} & \bar{\Phi}_{23} \\ \Phi_{13}^T & \bar{\Phi}_{23}^T & \Phi_{33} \end{bmatrix}, \bar{\Phi}_{22} = \tau^* G^T RG + \bar{\Omega},
$$

$$
\bar{\Phi}_{23} = -g(x)K^T P \otimes I_n,
$$

$$
\bar{\Omega} = (PA + A^T P - g(x)PKC - g(x)C^T K^T P) \otimes I_n.
$$

If the gain K is defined by $K = P^{-1}C^T$, we can obtain $\bar{\Phi} < 0$ from the condition $\Phi < 0$. Since the matrices Q, R, P are positive definite, we have $V(t) \geq 0$. It can be obtained that $J(t) = \int_0^t S(\tau)d\tau < 0$ if $S(t) < 0$ holds. Since

$$J(x(t)) = V(t) + \int_0^t (||y(\tau)||^2 - \gamma^2 ||\omega(\tau)||^2)d\tau < 0,$$

we get $||y(t)||^2 < \gamma^2 ||\omega(t)||^2$, thus, $||y(t)||_2 < \gamma ||\omega(t)||_2$ holds.

Thus, the composite system is robustly asymptotically stable and satisfies $||y(t)||_2 \leq \gamma ||\omega(t)||_2$. □

7.4 DISTURBANCE OBSERVERS FOR MULTI-AGENT SYSTEMS WITH SWITCHING TOPOLOGIES

Consider system (7.3) with switching topologies $\{G_s : s = \sigma(t) \in \Pi_0\}$, where $\Pi_0 \subset Z$ is a finite index set, and $\sigma(t)$ is a switching signal that determines the network topology is strongly connected. Under arbitrary switching signal, the switching system induced by (7.3) takes the following form:

$$\begin{cases} \dot{x}_i(t) = & \sum_{j \in N_i} a_{s,ij}(x_j(t-\tau) - x_i(t-\tau)) \\ & + u_i(t) + g(x_i)(d_i(t) + \omega_i(t)), \\ y_i(t) = & x_i(t), \end{cases} \tag{7.25}$$

where $s = \sigma(t) \in \Pi_0$, $a_{s,ij} \geq 0$ denotes the weighted value between node i and node j at switching signal s. Let L_s is the Laplacian matrix of the switching topology G_s.

Theorem 7.4

Consider multi-agent systems (7.25) with time delays and switching topologies of G_s strongly connected for each $s = \sigma(t) \in \Pi_0$. The multi-agent systems with disturbances and disturbance uncertainties. If there exist appropriate positive definite matrices Q, R, P, constant number $\tau^* > 0$ and given parameters $\gamma > 0$, such that

$$\Phi = \begin{bmatrix} \Phi_{11} & \Phi_{12} & \Phi_{13} \\ \Phi_{12}^T & \Phi_{22} & \Phi_{23} \\ \Phi_{13}^T & \Phi_{23}^T & \Phi_{33} \end{bmatrix} < 0, \tag{7.26}$$

where

$$\Phi_{11} = \begin{bmatrix} \bar{\Phi}_{11} & 0 \\ 0 & -Q + \tau^* \bar{L}_s^T R \bar{L}_s \end{bmatrix},$$

$$\bar{\Phi}_{11} = -2\bar{L}_s + Q + \tau^* \bar{L}_s^T R^{-1} \bar{L}_s + I_n, \Phi_{12} = [G^T, -\tau^* G^T R^T \bar{L}_s^T]^T,$$

$$\Phi_{22} = \tau^* G^T RG + \Omega, \Phi_{13} = [H^T, -\tau^* H^T R^T \bar{L}_s^T]^T, , \Phi_{23} = -G$$

$$\Phi_{33} = \tau^* H^T RH - \gamma^2 I_n, \Omega = (A^T P + PA - 2g(x)C^T C) \otimes I_n,$$

$$H = g(x) \otimes I_n, G = g(x)C \otimes I_n, \bar{L}_s = L_s + B.$$

Then, the closed-loop dynamical networks under the disturbance observers (7.11) can reach expected consensus asymptotically for $\tau \in (0, \tau^*)$ in the absence of distance uncertainties and satisfies $||y(t)||_2 \leq \gamma ||\omega(t)||_2$, where

$$y(t) = [y_1(t), ..., y_n(t)]^T, y_i(t) = \delta x_i(t), \omega(t) = [\omega_1(t), ..., \omega_n(t)]^T$$

■

Proof. The proof is similar to that of Theorem 7.3.   □

7.5 EXAMPLE STUDY

Consider a multi-agent network with switching topologies (Figure 7.2). The weighted values of the graph are given randomly in $[0, 1]$ and Laplacian matrix $L_{\sigma(t)}$ can be achieved depending on the adjacency matrix of switching topologies.

Case 1.

Suppose there do not exist disturbances in multi-agent systems (or there exist measurable disturbances). Let the initial states of agents be generated randomly in $[0, 6]$, and the expected consensus track be $\bar{x} = 0.5$. We apply the pinning control (7.7) with a small fraction $\delta = 0.20$ of the nodes, i.e., only one node is pinned. Suppose the control gain $b_1 = 1$, the critical value of the time delay is $\tau^* = 0.5476$ s by calculating the linear matrix inequalities (LMIs) from Theorem 7.1. The moving states of multi-agent systems without disturbances are plotted in Figure 7.3 with time delay $\tau = 0.5$ s. We can see that the consensus track $\bar{x} = 0.5$ is reached asymptotically.

Case 2.

Consider multi-agent systems with disturbances and disturbance uncertainties $\omega_i(t) = 0$. Suppose the parameters of the exogenous disturbance system (7.4) are

$$A = \begin{bmatrix} 0 & 2 \\ -2 & 0 \end{bmatrix}, C = [1, 0]$$

with initialized value $\xi_i = [0.5 \sin 1, 0.5 \cos 1]^T$. In this example, we suppose $g(x) = 1$. By applying linear matrix inequalities (LMIs) packages of MATLAB, we get

$$P = \begin{bmatrix} \frac{1}{2} & -\frac{1}{4} \\ -\frac{1}{4} & \frac{1}{3} \end{bmatrix}$$

After P is determined, the gain K can be calculated by $K = P^{-1}C^{T}$. In Figure 7.4, the outputs of the multi-agent systems with exogenous disturbances are shown. The estimation given by observer (7.11) is shown in Figure 7.5, where the dot line is the estimation of the exogenous disturbance. We can see the observer exhibits excellent tracking performance. In Figure 7.6, the moving states of multi-agent systems with DOBC are plotted, it is shown that the expected consensus state has been achieved for multi-agent systems with the exogenous disturbances under DOBC.

Case 3.

Consider the multi-agent systems with disturbance parameters $g(x) = 1 + x^2$ and disturbance uncertainties $\omega_i(t) = 0$. Suppose the exogenous disturbance system is the same as that in Case 2. Applying the same simulating procedure, the moving states of multi-agent systems with DOBC are plotted in Figure 7.7. The expected consensus state can be achieved for delayed multi-agent systems with the exogenous disturbances under DOBC.

Case 4.

Consider multi-agent systems with disturbance parameters $g(x) = 1 + x^2$ and disturbance uncertainties $\omega_i(t) = 1$. Suppose the exogenous disturbance system is the same as that in Case 2. By applying LMIs packages of MATLAB, we can obtain

$$P^{-1} = \begin{bmatrix} 3 & 3 \\ 3 & 4 \end{bmatrix}, \ K = [3,3]^T$$

Applying the same simulating procedure as in Case 3, the moving states of multi-agent systems with DOBC are plotted in Figure 7.8. The expected consensus state can be achieved for delayed multi-agent systems with the exogenous disturbances under DOBC.

7.6 CONCLUSION

The problem of cooperation in a team of unmanned multi-agent systems with time delays is considered under the effects of disturbances. A pinning control strategy is designed for a part of the agents of a multi-agent system without disturbances, where the control can bring multiple agents' states to an expected consensus track. Under the effects of the disturbances, nonlinear disturbance observers are developed by an exogenous system to estimate the disturbances. The disturbance observers are integrated with the controller by replacing the disturbance in the control law with its estimation yielded by the disturbance observer. Furthermore, asymptotical consensus of the multi-agent systems with disturbances under the composite controller is achieved.

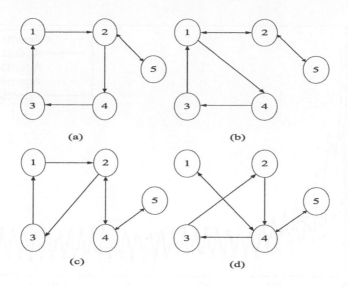

FIGURE 7.2   Switching graph of multi-agent systems.

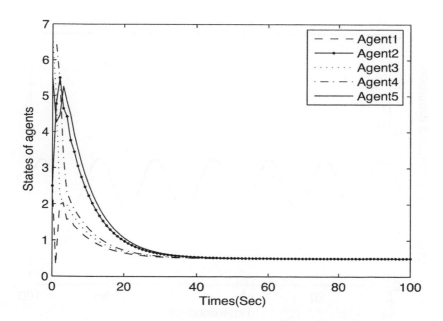

FIGURE 7.3   States of multi-agent systems without disturbance on pinning control.

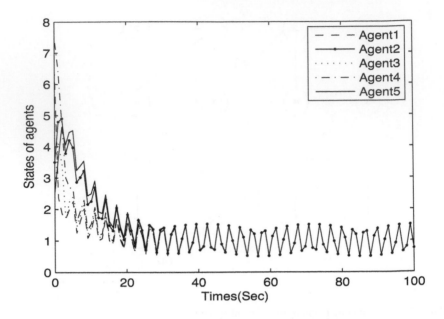

FIGURE 7.4   States of multi-agent systems with exogenous disturbances.

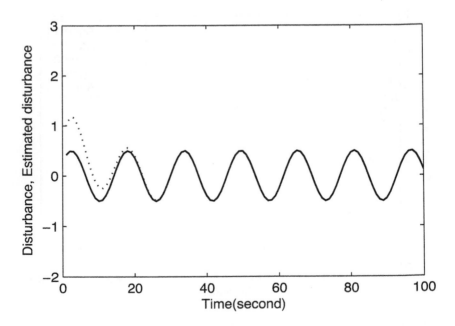

FIGURE 7.5   Disturbances estimated by DOBC.

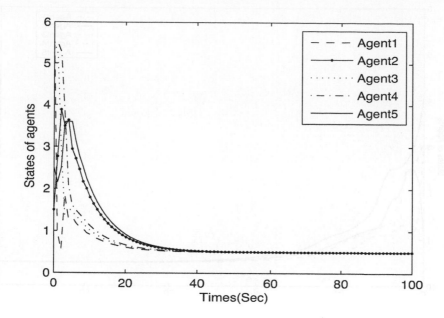

FIGURE 7.6 States of multi-agent systems with DOBC.

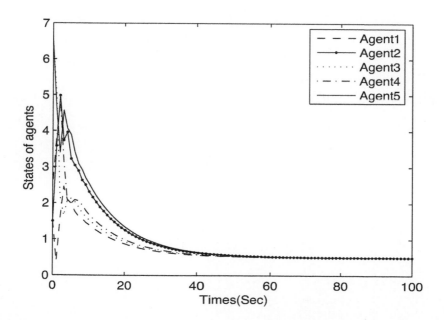

FIGURE 7.7 States of multi-agent systems with DOBC and $g(x) = 1 + x^2$.

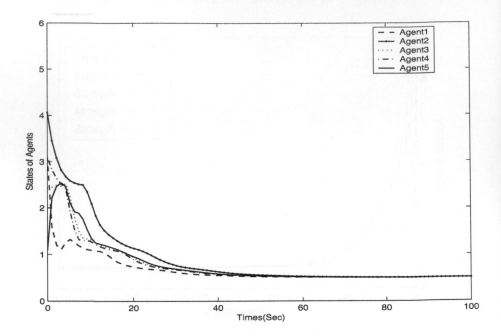

FIGURE 7.8   States of multi-agent systems with DOBC and disturbance uncertainties.

8 Hierarchical Anti-Disturbance Adaptive Controller Design

8.1 INTRODUCTION

In general, there are two kinds of methodologies focusing on the anti-disturbance control problem for nonlinear systems with unknown disturbances. The first one is the disturbance attenuation approach such as nonlinear H_∞ control theory feasible for norm bounded disturbances [7, 80]. Another is the disturbance rejection method including nonlinear output regulation theory which has been shown to be good performance on the disturbance described by exogenous models (see, e.g., [47, 106, 126, 159, 180, 188]). A disturbance observer based control (DOBC) approach has been widely used in many practical engineering contexts, such as robot control, table drive systems, hard disks, active magnetic bearings, vibrational microelectronic mechanical system (MEMS) gyroscopes (see survey in [87] and references therein). In the first stage, only linear DOBC is involved in the frequency domain, where the nonlinearities and uncertainties are treated as a disturbance to be estimated and rejected [10]. In [85], DOBC was first generalized to uncertain systems with an exogenous disturbance model in the time-domain formulations. The nonlinear DOBC law was proposed for robots [32] with harmonic disturbances. Several valuable new related results have been given for nonlinear systems with special structures [31, 182, 221].

Generally speaking, there are several drawbacks to be overcome on nonlinear anti-disturbance control. Firstly, in practical engineering, the disturbance may originate from different aspects and can be described by a composite form rather than a single variable. In this case, the H_∞ control may be too conservative to provide a highly accurate control performance. On the other hand, disturbance rejection approaches usually need precise models for both the controlled system and the disturbance system. This confines their applications since the disturbance is also described by a single output variable from a precise exosystem and robustness is difficult to guarantee. For example, although DOBC is a valid disturbance rejection strategy for nonlinear systems with harmonic disturbances, the system performance will deteriorate if the disturbance model cannot be described precisely [85].

In this chapter the disturbance is formulated as a composite one with two different parts. The first part is described by an exogenous model, while the second one is uncertain but bounded by a given function. A new hierarchical anti-disturbance control scheme is presented, where a nonlinear disturbance observer (NLDO) for the modeled disturbance is designed and integrated separately with an adaptive controller to reject the unknown composite disturbances. It is shown that with the proposed controller the closed-loop system is uniformly ultimately bounded (UUB) or practically stable (see [15, 205]). Finally, the approach is applied to a missile control system. Simulation indicates that improved performance is obtained in comparison with the results in [31].

8.2 PROBLEM FORMULATION AND PRELIMINARIES

Consider a nonlinear system described by

$$\dot{x} = F(x) + g_1(x)v + g_2(x)d \tag{8.1}$$

where $x \in R^n$, the functions $F(x)$, $g_1(x)$ and $g_2(x)$ are smooth functions and assumed to be bounded with respect to x. x, v and d are the state, input and unknown disturbance, respectively.

In this chapter, we suppose d is a composite or multiple disturbance as

$$d = d_h + d_\theta.$$

d_h is a dynamic disturbance described by the following model but with unknown initial conditions

$$\begin{cases} \dot{w} = Aw \\ d_h = Cw \end{cases} \tag{8.2}$$

where (A, C) is assumed to be observable. It is noted that the well-known harmonic disturbances with known frequency but unknown amplitude and phase are special cases of this model. $d_\theta = \xi(x, w, t)$ is supposed to represent a kind of bounded disturbances. $\xi(x, w, t)$ satisfies Lipschitz condition

$$\|\xi(x, w_1, t) - \xi(x, w_2, t)\| \leq \eta \|w_1 - w_2\| \tag{8.3}$$

and the bounded condition

$$\|\xi(x, w, t)\| \leq \rho(x, w, t)\theta^* \tag{8.4}$$

where $\rho(x, w, t)$ is continuous and θ^* is an unknown constant to be determined below.

Thus, the original control system can also be described by

$$\dot{x} = f(x) + g_1(x)u + g_2(x)(d_h + d_\theta). \tag{8.5}$$

The key problem of DOBC control is to estimate the disturbance d_h and then integrate the observer with a rejection controller. For composite disturbance d, a composite anti-disturbance control law needs to be established with a hierarchical structure. The problem to be considered in this chapter includes three parts:

- To estimate and reject the disturbances d_h in (8.1) using a nonlinear observer and guarantee the convergence with a prespecified accuracy;

- To design a nonlinear control to compensate for the uncertain signal d_θ in (8.1) using a nonlinear adaptive controller;
- To integrate the observer and compensator with a conventional control for rejecting the composite disturbances and guaranteeing the stability of the closed-loop system.

8.3 NONLINEAR DISTURBANCE OBSERVER

To reject the disturbances, the following assumption is needed in this chapter, which can be satisfied in many practical problems (see [32]).

Assumption 8.1 *There exists a smooth function $h(x)$ and an integer $\rho \geq 1$ such that $n(x) = L_{g_2} L_f^{\rho-1} h(x) \neq 0$ and $n(x)$ is bounded with respect to x in its operation region.*

A nonlinear disturbance observer for d_h is proposed as

$$
\begin{cases}
\dot{z} &= A[z + p(x)] - l(x)g_2(x)Cz \\
&\quad -l(x)[g_2(x)Cp(x) + f(x) + g_1(x)u] \\
\widehat{w} &= z + p(x) \\
\widehat{d_h} &= C\widehat{w}
\end{cases}
\tag{8.6}
$$

where $p(x)$ is a nonlinear function to be designed and the nonlinear observer gain $l(x)$ is defined as

$$
l(x) = \frac{\partial p(x)}{\partial x}
\tag{8.7}
$$

Define $e = w - \widehat{w}$. Under (8.2), (8.5), (8.6) and (8.7), it can be verified that

$$
\begin{aligned}
\dot{e} &= \dot{w} - \dot{\widehat{w}} = Aw - \dot{z} - \frac{\partial p(x)}{\partial x}\dot{x} \\
&= Aw - \{Az - l(x)g_2(x)Cz + Ap(x) \\
&\quad -l(x)[g_2(x)Cp(x) + f(x) + g_1(x)u]\} \\
&\quad -l(x)[f(x) + g_1(x)u + g_2(x)(d_h + d_\theta)] \\
&= Aw - [A - l(x)g_2(x)C][\widehat{w} - p(x)] \\
&\quad -Ap(x) + l(x)[g_2(x)(Cp(x) - Cw - d_\theta)] \\
&= Ae - l(x)g_2(x)Ce - l(x)g_2(x)d_\theta.
\end{aligned}
$$

Our first objective is to find a continuous and bounded function $l(x)$ satisfying (8.7) such that the uncertain nonlinear system

$$
\dot{e} = (A - l(x)g_2(x)C)e - l(x)g_2(x)d_\theta
\tag{8.8}
$$

is practically stable.

Under Assumption 8.1, if we select

$$l(x) = \begin{bmatrix} l_1 \\ l_2 \end{bmatrix} \frac{\partial L_f^{\rho-1}h(x)}{\partial x} \tag{8.9}$$

then we have

$$l(x)g_2(x) = \begin{bmatrix} l_1 \\ l_2 \end{bmatrix} n(x)$$

where

$$n(x) = L_{g_2}L_f^{\rho-1}h(x)$$

In this case, we can find a constant n_0 and a bounded nonlinear function $n_1(x)$ satisfying

$$n_1^2(x) \leq \bar{n}_1^2 \tag{8.10}$$

for all x in operation domain such that

$$n(x) = -n_0 + n_1(x) \tag{8.11}$$

For shorthand description, we denote $0 < n^2(x) \leq \bar{n}^2$, where $\bar{n}_1^2$, $\bar{n}^2$ are known constants. Thus, (8.8) can be rewritten to

$$\dot{e} = [\bar{A} - Ln_1(x)C]e - Ln(x)d_\theta \tag{8.12}$$

where

$$\bar{A} = A + n_0LC, L = \begin{bmatrix} l_1 \\ l_2 \end{bmatrix}$$

The following results provide a design scheme for the disturbance observer (see (8.6) and (8.9)) with the guaranteed stability of (8.12) or (8.8).

Theorem 8.1

Under Assumption 8.1, if there exist $P \in R^{2\times 2} > 0$ and $Q \in R^{2\times 1}$ satisfying

$$\begin{bmatrix} A^TP + PA + n_0QC + n_0C^TQ^T + C^TC & Q \\ Q^T & -(\bar{n}_1^2 + \beta^2\bar{n}^2)^{-1}I \end{bmatrix} < 0, \tag{8.13}$$

where $L = P^{-1}Q$, then the system (8.12) is UUB for all x varying in the operating region.   ∎

   Proof. Firstly (8.13) implies

$$A^TP + PA + n_0QC + n_0C^TQ^T + C^TC \\ + (\bar{n}_1^2 + \beta^2\bar{n}^2)QQ^T + \alpha_0I < 0 \tag{8.14}$$

where $\alpha_0 > 0$ is a given constant.

Denoting $W(e) = e^T Pe$, by (8.12) under the case of $d_\theta = 0$ we have

$$\dot{W} = e^T \left\{ [\bar{A} - Ln_1(x)C]^T P + P[\bar{A} - Ln_1(x)C] \right\} e$$

Along with (8.12) we can get

$$
\begin{aligned}
\dot{W} &= e^T \left\{ [\bar{A} - Ln_1(x)C]^T P + P[\bar{A} - Ln_1(x)C] \right\} e \\
&\quad - 2e^T PLn(x)d_\theta \\
&= e^T \left\{ \bar{A}^T P + P\bar{A} + Qn_1^2 Q^T + C^T C \right. \\
&\quad \left. - [Qn_1(x) + C^T][n_1(x)Q^T + C] \right\} e - 2e^T Qn(x)d_\theta \\
&\leq e^T \left\{ \bar{A}^T P + P\bar{A} + \bar{n}_1^2 QQ^T + C^T C \right\} e - 2e^T Qn(x)d_\theta \\
&\leq e^T \left\{ \bar{A}^T P + P\bar{A} + (\bar{n}_1^2 + \beta^2 \bar{n}^2)QQ^T + C^T C \right\} e + \beta^{-2}\bar{\xi}^2 \\
&\leq -\alpha_0 \|e\|^2 + \beta^{-2}\bar{\xi}^2 \\
&\leq -\alpha_0 \sigma_{\max}^{-1}(P)W + \beta^{-2}\bar{\xi}^2
\end{aligned}
$$

From the definition of W, we have

$$\|e(t)\|^2 \leq \|e(0)\|^2 \exp(-\alpha_0 \sigma_{\max}^{-1}(P)t) + \varepsilon, \tag{8.15}$$

where

$$\varepsilon = \frac{\sigma_{\max}(P)\bar{\xi}^2}{\alpha_0 \beta^2 \sigma_{\min}(P)}.$$

It means the system (8.6) is UUB, and the convergent rate of the state can be adjusted by selecting the parameter β and α_0 without respect to x. $\square$

8.4 STABILITY OF THE DOBC SYSTEMS

In this section the estimation of d_h will be used to reject it and the stability will be studied. Similarly to [201], the following two assumptions are needed in this chapter to guarantee the DOBC system is wellposed and globally stable.

Assumption 8.2 *For system (8.1), there exists a stabilizing law $v = k(x) + u$ so that $x = 0$ is a uniform globally asymptotically stable equilibrium of $\dot{x} = f(x)$, where*

$$f(x) = F(x) + g_1(x)k(x)$$

Moreover, there is a function $V_0(x) : R_+ \times R^n \to R_+$ satisfying

$$\gamma_1(\|x\|) \leq V_0(x) \leq \gamma_2(\|x\|), \frac{\partial^T V_0(x)}{\partial x} f(x) \leq -\gamma_3(\|x\|),$$

and

$$\lim_{\|x\| \to \infty} \left\| \frac{\partial^T V_0(x)}{\partial x} \right\| / \gamma_3(\|x\|) = 0,$$

where $\gamma_1(\cdot)$, $\gamma_2(\cdot)$ are of $K_\infty - class$ and $\gamma_3(\cdot)$ of $K - class$.

Assumption 8.3 *There exists a bounded function $g_0(x)$ such that*

$$g_2(x) = g_1(x)g_0(x)$$

Denote $u = u_1 + u_2$ as the hierarchical control law, where

$$u_1 = -g_0(x)\widehat{d}_h.$$

Equation (8.5) becomes

$$\dot{x} = f(x) + g_2(x)Ce + g_1(x)u_2 + g_2(x)d_\theta \tag{8.16}$$

Rewrite the composite system consisting of (8.16) and (8.8) into

$$\begin{cases} \dot{x} &= G(x, w, \widehat{w}, t) + g_1(x)[u_2 + g_0(x)\xi(x, \widehat{w}, t)] \\ \dot{e} &= [A - l(x)g_2(x)C]e - Ln(x)d_\theta \end{cases} \tag{8.17}$$

where

$$G(x, w, \widehat{w}, t) = f(x) + g_2(x)Ce + g_1(x)g_0(x)[\xi(x, w, t) - \xi(x, \widehat{w}, t)] \tag{8.18}$$

Firstly we only need to consider the following auxiliary system

$$\begin{cases} \dot{x} &= G(x, w, \widehat{w}, t) \\ \dot{e} &= [A - l(x)g_2(x)C]e - Ln(x)d_\theta \end{cases} \tag{8.19}$$

Theorem 8.2

For the auxiliary system (8.19) under Assumptions 8.1-8.3, if there exist $P \in R^{2\times 2} > 0$ and $Q \in R^{2\times 1}$ satisfying (8.13), then (8.19) is uniformly ultimately bounded. ∎

Proof. Along the trajectories of (8.19) we can get

$$
\begin{aligned}
\frac{dV_0(x)}{dt} &= \frac{\partial^T V_0(x)}{\partial x} G(x, w, \widehat{w}, t) \\[2mm]
&= \frac{\partial^T V_0(x)}{\partial x} f(x) + \frac{\partial^T V_0(x)}{\partial x} g_2(x) Ce \\[2mm]
&\quad + \frac{\partial^T V_0(x)}{\partial x} g_1(x) g_0(x) [\xi(x, w, t) - \xi(x, \widehat{w}, t)] \\[2mm]
&\leq -\gamma_3(\|x\|) + \left\| \frac{\partial^T V_0(x)}{\partial x} \right\| \|g_2(x)\| \|C\| \|e\| \\[2mm]
&\quad + \eta \left\| \frac{\partial^T V_0(x)}{\partial x} \right\| \|g_1(x)\| \|g_0(x)\| \|e\|
\end{aligned}
\tag{8.20}
$$

Based on Assumption 8.2 and (8.15), there exists $\sigma > 0$ such that whenever $\|x\| \geq \sigma$,

$$
\frac{\left\| \frac{\partial^T V_0(x)}{\partial x} \right\|}{\gamma_3(\|x\|)} \|g_2(x)\| \|C\| \|e\| < \frac{\kappa}{2} < 1/2,
$$

$$
\eta \frac{\left\| \frac{\partial^T V_0(x)}{\partial x} \right\|}{\gamma_3(\|x\|)} \|g_1(x)\| \|g_0(x)\| \|e\| < \frac{\kappa}{2} < 1/2
$$

Thus, we have

$$
\frac{dV_0(x)}{dt} \leq (-1 + \kappa)\gamma_3(\|x\|),
\tag{8.21}
$$

whenever $\|x\| \geq \sigma$.

From (8.15) for all $t \geq 0$ we can obtain

$$
\|x(t)\| \leq \max\left\{ \|x(0)\|, \frac{1}{(1-\kappa)} \gamma_3^{-1}(\sigma) \right\} := \delta
$$

which implies that x is uniformly bounded. Letting

$$
v := \sup_x \left\{ \left\| \frac{\partial^T V_0(x)}{\partial x} \right\| \|g_2(x)\| \|C\| + \eta \left\| \frac{\partial^T V_0(x)}{\partial x} \right\| \|g_1(x)\| \|g_0(x)\| \right\}
$$

yields

$$
\frac{dV_0(x)}{dt} \leq -\gamma_3(\|x\|) + v \|e\|
\tag{8.22}
$$

For the closed-loop system (8.19), denoting $V_1(x,e) = V_0(x) + W(e)$, based on (8.15), (8.20) and (8.22), we have

$$
\begin{aligned}
\frac{dV_1(x,e)}{dt} &\leq -\gamma_3(\|x\|) + v\|e\| - \alpha_0\|e\|^2 + \beta^{-2}\overline{\xi}^2 \\
&\leq -\gamma_3(\|x\|) - \frac{1}{2}\alpha_0\|e\|^2 + \varepsilon \\
&\leq -\gamma_4\left(\left\|\begin{bmatrix} x \\ e \end{bmatrix}\right\|\right) + \varepsilon
\end{aligned}
$$
(8.23)

where $\gamma_4(\cdot)$ is also of $K - class$ respective to x and e and

$$
\varepsilon := \alpha_0^{-1}v^2 + \beta^{-2}\overline{\xi}^2.
$$

Thus, we can conclude that (8.19) is also UUB.　　　　　　　　□

8.5 HIERARCHICAL ADAPTIVE CONTROL FOR THE COMPOSITE SYSTEMS

In the following we aim to find u_2 in the hierarchical controller as well as in (8.17) to reject d_θ.

Define

$$
\mu(x,t) = g_1^T(x)\frac{\partial V_0(x)}{\partial x}
$$

and

$$
u_2 = \varphi(x,\widehat{w},t) = -\frac{\|\mu(x,t)\|[\rho(x,\widehat{w},t)\widehat{\theta}]^2\|g_0(x)\|^2}{\|\mu(x,t)\|\|g_0(x)\|\|\rho(x,\widehat{w},t)\widehat{\theta}\| + \zeta}
$$
(8.24)

where ζ is any positive constant, $\widehat{\theta}$ is the estimate of the unknown parameter vector θ^* (see [201]).

$\widehat{\theta}$ is updated by the following adaptive law

$$
\frac{d\widehat{\theta}}{dt} = -\delta_1\widehat{\theta} + \frac{\delta_2}{2}\|\mu(x,t)\|\|g_0(x)\|\rho(x,\widehat{w},t)
$$
(8.25)

where $\delta_1 > 0$, $\delta_2 > 0$ and $\widehat{\theta}(0)$ is finite and given. Denoting $\widetilde{\theta} = \widehat{\theta} - \theta^*$, we have

$$
\frac{d\widetilde{\theta}}{dt} = -\delta_1\widetilde{\theta} + \frac{\delta_2}{2}\|\mu(x,t)\|\|g_0(x)\|\rho(x,\widehat{w},t) - \delta_1\theta^*.
$$
(8.26)

Theorem 8.3

For the composite closed-loop system (8.17), under Assumptions 8.1-8.3, if there exist $P \in R^{2\times2} > 0$ and $Q \in R^{2\times1}$ satisfying (8.13), then (8.17) with adaptive laws (8.22) and (8.24) is UUB.　　■

Proof. Define a Lyapunov function candidate as

$$V(x, e, \widetilde{\theta}) = V_0(x) + W(e) + \delta_2^{-1} \widetilde{\theta}^T \widetilde{\theta}.$$

Take the derivative of $V(\cdot)$ along the composite closed-loop system leads to

$$\frac{dV(x, e, \widetilde{\theta})}{dt} = \frac{\partial^T V_0(x)}{\partial x} G(x, w, \widehat{w}, t) + \frac{\partial^T V_0(x)}{\partial x} g_1(x) [\varphi(x, \widehat{w}, t)$$

$$+ g_0(x) \xi(x, \widehat{w}, t)] + 2\delta_2^{-1} \widetilde{\theta} \frac{d\widetilde{\theta}}{dt} + \dot{W}$$

Based on (8.21), (8.22) and (8.24), we have ($\delta_3 = \delta_1 \delta_2^{-1}$)

$$\frac{dV(x, \widetilde{\theta})}{dt} \leq -\gamma_4\left(\left\|\begin{bmatrix} x \\ e \end{bmatrix}\right\|\right) + \varepsilon + \mu(x, t) [\varphi(x, \widehat{w}, t)$$

$$+ g_0(x) \xi(x, \widehat{w}, t)] - 2\delta_3 \widetilde{\theta}^T \widetilde{\theta} +$$

$$\widetilde{\theta} \|\mu(x, t)\| \|g_0(x)\| \rho(x, \widehat{w}, t) - 2\delta_3 \widetilde{\theta}^T \theta^*$$

$$\leq -\gamma_4\left(\left\|\begin{bmatrix} x \\ e \end{bmatrix}\right\|\right) + \varepsilon + \mu(x, t) \varphi(x, \widehat{w}, t)$$

$$+ \|\mu(x, t)\| \|g_0(x)\| \rho(x, \widehat{w}, t) \widehat{\theta}$$

$$- 2\delta_3 \widetilde{\theta}^T \widetilde{\theta} - 2\delta_3 \widetilde{\theta}^T \theta^*.$$

With (8.22), we get

$$\mu(x, t) \varphi(x, \widehat{w}, t) + \|\mu(x, t)\| \|g_0(x)\| \rho(x, \widehat{w}, t) \widehat{\theta}$$

$$\leq \frac{\|\mu(x, t)\| \|g_0(x)\| \|\rho(x, \widehat{w}, t) \widehat{\theta}\| \zeta}{\|\mu(x, t)\| \|g_0(x)\| \|\rho(x, \widehat{w}, t) \widehat{\theta}\| + \zeta} \leq \zeta$$

On the other hand, it is noted that

$$-2\delta_3 \widetilde{\theta}^T \widetilde{\theta} - 2\delta_3 \widetilde{\theta}^T \theta^* \leq -\delta_3 \left\|\widetilde{\theta}\right\|^2 + \delta_3 \|\theta^*\|^2$$

then we can obtain

$$\frac{dV(x, \widetilde{\theta})}{dt} \leq -\gamma_4\left(\left\|\begin{bmatrix} x \\ e \end{bmatrix}\right\|\right) + \varepsilon - \delta_3 \left\|\widetilde{\theta}\right\|^2 + \zeta + \delta_3 \|\theta^*\|^2$$

$$\leq -\gamma_5\left(\left\|\begin{bmatrix} x^T & e^T & \widetilde{\theta} \end{bmatrix}^T\right\|\right) + \varsigma \qquad (8.27)$$

where $\gamma_5(\cdot)$ is of $K - class$ and

$$\varsigma = \varepsilon + \zeta + \delta_3 \|\theta^*\|^2.$$

(8.25) implies that the composite closed-loop system is UUB.   □

Remark 8.1 Since ε can be designed sufficiently small by adjusting β and α_0 via P, and ζ and δ_3 can also be selected sufficiently small in (8.23) and (8.22), every system response can be confined within a sufficiently small neighborhood of the zero state. However, this may result in high gain of the controllers.

Based on Theorems 8.1 and 8.3, the design method can be summarized as follows:

- Design the nonlinear observer as (8.6) with (8.12) based on Theorem 8.1 to estimate the disturbances.
- Construct the adaptive controller u_2 as (8.24) with adaptive law (8.25).
- Integrate the hierarchical controller.

$$u = -g_0(x)\widehat{d}_h + u_2$$

8.6 SIMULATION EXAMPLES

As in [31], the longitudinal dynamics of a missile can be described by

$$\dot{\alpha} = f_1(\alpha) + q + b_1(\alpha)\delta \tag{8.28}$$

and

$$\dot{q} = f_2(\alpha) + b_2\delta \tag{8.29}$$

where α is the angle of attack (deg), q the pitch rate (deg/s), and δ the tail fin deflection (deg). The disturbances d represent all disturbance torques that may be caused by unmodelled dynamics, external wind, and the variation of aerodynamic coefficients, etc. The nonlinear functions $f_1(\alpha)$, $f_1(\alpha)$, $b_1(\alpha)$ and b_2 are determined by the aerodynamic coefficients. For instance, when the missile travels at Mach 3 at an altitude of $6{,}095m$ $(20{,}000foot)$ and the angle of attack $|\alpha| \le 20$ deg, they are given by

$$f_1(\alpha) \quad = \quad \frac{180gQS}{\pi WV}\cos(\pi\alpha/180)(1.03 \times 10^{-4}\alpha^3$$

$$-9.45 \times 10^{-3}\alpha|\alpha| - 1.7 \times 10^{-1}\alpha) \tag{8.30}$$

$$f_2(\alpha) \quad = \quad \frac{180QSI_d}{I_{yy}}(2.15 \times 10^{-4}\alpha^3 -$$

$$1.95 \times 10^{-2}\alpha|\alpha| + 5.1 \times 10^{-2}\alpha), \tag{8.31}$$

$$b_1(\alpha) = -3.4 \times 10^{-2}\frac{180gQS}{\pi WV}\cos(\pi\alpha/180), \tag{8.32}$$

and

$$b_2 = -0.206 \frac{180 Q S I_d}{I_{yy}}. \tag{8.33}$$

The tail fin actuator dynamics are approximated by a first-order lag

$$\dot{\delta} = \frac{1}{t_1}(-\delta + v) + d \tag{8.34}$$

where v is the commanded fin deflection (deg) and t_1 the time constant. In [31], only unknown slowly time-varying disturbance is considered. In fact, the tail fin actuator will submit to exogenous disturbance caused by airstream vibration, we model this disturbance as an action of harmonic and frequency uncertain disturbance which widely exists in aerospace engineering. So d in this chapter has form of $d = d_h + d_\theta$, where the meanings of d_h and d_θ are as mentioned in Section 8.2. Thus, a new disturbance estimation and control strategy is needed to enhance the control performance further. The significance of the parameters in (8.30)-(8.33) and their values for the missile under consideration are listed in Table 8.1. In the absence of the disturbance d, an auto-pilot for the missile to track a reference angle of attack $w(t)$ can be designed using a nonlinear dynamic inverse control. The resultant control law is given by

$$u_0 = \delta - \frac{t_1}{b_1}[k_1(\alpha - w(t))$$

$$+ k_2(f_1 + q + b_1(\alpha)\delta - \dot{w}) + m - \ddot{w}] \tag{8.35}$$

where

$$m = (\frac{\partial f_1(\alpha)}{\partial \alpha} + \frac{\partial b_1(\alpha)}{\partial \alpha}\delta)(f_1(\alpha) + q + b_1(\alpha)\delta)$$

$$+ f_2(\alpha) + b_2\delta \tag{8.36}$$

and k_1 and k_2 are constant gains to be designed according to desired closed-loop behaviors.

The longitudinal dynamics of the missile can be put into the general description of an affine system as (8.1) with

$$F(x) = \begin{bmatrix} f_1(\alpha) + q + b_1(\alpha)\delta \\ f_2(\alpha) + b_2\delta \\ -\frac{1}{t_1}\delta \end{bmatrix}, g_1(x) = \begin{bmatrix} 0 \\ 0 \\ \frac{1}{t_1} \end{bmatrix}$$

$$g_2(x) = \begin{bmatrix} 0 \\ 0 \\ 1 \end{bmatrix}, x = \begin{bmatrix} \alpha \\ q \\ \delta \end{bmatrix} \tag{8.37}$$

Since $g_2(x)$ is a constant vector, from (8.9) and (8.11), we can choose

$$l(x)g_2(x) = Ln(x) = \begin{bmatrix} l_1 \\ l_2 \end{bmatrix} n(x), n(x) = b_1(\alpha).$$

TABLE 8.1

The parameters in longitudinal dynamics of the missile.

Parameter	Symbol	Value
Weight	W	$4410\ kg$
Velocity	V	$947.6\ m/s$
Pitch moment of inertia	I_{yy}	$247.44\ kg$
Dynamic pressure	Q	$293,638\ M/m^2$
Reference area	S	$0.04087\ m^2$
Reference diameter	I_d	$0.229\ m$
Gravitational acceleration	g	$9.8\ m/s^2$
Time constant of tin actuator	t_1	$0.1\ s$

Source: Reference [30].

Select $\beta^2 = 0.25$, correspondingly, $\bar{n}_1 = 0.5$ and $\bar{n} = 1$, then (8.13) yields

$$A^T P + PA + QC + C^T Q^T + C^T C + 0.5 QQ^T < 0,$$

with

$$A = \begin{bmatrix} 0 & 0.25 \\ -0.25 & 0 \end{bmatrix}, C = \begin{bmatrix} 10 & 0 \end{bmatrix}$$

from which we can obtain

$$L = P^{-1}Q = \begin{bmatrix} -0.1657 \\ 0.0727 \end{bmatrix}$$

and $\alpha_0 \leq 12.95$ in (8.14). Corresponding to (8.24), we have

$$\mu(x,t) = [0\ 0\ 1]\frac{\partial V_0(x)}{\partial x}$$

Figure 8.1 shows that although in the presence of uncertain frequency d_θ, our disturbance observer could estimate the harmonic disturbance effectively, and the estimate error may be confined to a sufficiently small region. In Figure 8.2, the hierarchical controller is applied to the missile system, it is clear that the system UUB could be guaranteed. Also, the good track property can be obtained by a proper selection of observer or controller parameters according to (8.13), (8.24) and (8.35) as shown in Figure 8.3. The solid line is the output using our controller based on NLDO, the disturbance including harmonic and uncertainty is rejected to a large extent. The output using the controller presented in [31] is plotted by a dotted line, from which the angle of attack will vibrate slightly caused by disturbance estimation error.

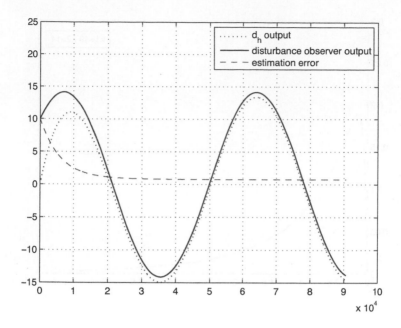

FIGURE 8.1   The comparison of d_h and disturbance observer.

8.7   CONCLUSION

In this chapter, the anti-disturbance control problem is considered for a class of nonlinear systems with multiple disturbances. The disturbance is formulated as a composite one with two different variables. The first one is described by an exogenous model and the second one is uncertain but bounded by a given function. A new nonlinear disturbance observer is designed to estimate the composite disturbances. Then, a hierarchical control strategy consisting of DOBC and a robust adaptive controller is presented to achieve anti-disturbance performance. Stability analysis for both the error estimation systems and the composite closed-loop system is provided to show that stability is guaranteed. Finally, we apply this approach into a missile control system. It is shown that with the hierarchical anti-disturbance controller, improved robustness and satisfactory performance can be obtained.

It is noted that the proposed hierarchical anti-disturbance controller has a flexible structure and is easily combined with other controllers to deal with various types of disturbances (see [87] for details). Another potential application is to use the disturbance observer to study fault diagnostics and tolerance control problems (see [70, 72, 231]).

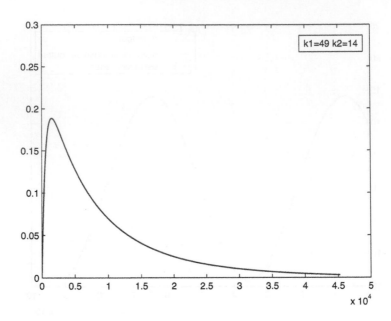

FIGURE 8.2   The missile system UUB property based on NLDO.

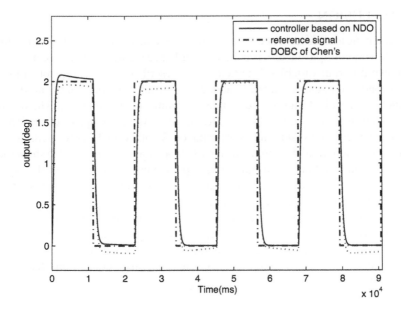

FIGURE 8.3   Track property of missile system based on NLDO.

9 Estimation and Rejection of Unknown Sinusoidal Disturbances Using TSNLDO

9.1 INTRODUCTION

Over the last two decades, the rejection problem of unknown neutral stable disturbances for nonlinear systems has attracted considerable attention. A prevalent control strategy is the well-known output regulation theory, where the unknown disturbances can be treated as an autonomous exosystem ([12, 106, 112]). The compensation of the disturbances can be realized in the feedback path with the celebrated internal model principle. If the parameter is unknown, the situation becomes complex. For linear systems, the adaptive regulator is designed in [145] which tunes the internal model of the unknown exosystem so that it will asymptotically reproduce the unknown exosystem and the corresponding open-loop control signal. In [16], periodic disturbances with uncertain frequency are cancelled using an adaptive internal model based on singular perturbation and averaging theories, and local exponential stability can be obtained. The more general case where there exists an uncertain dynamic in a system model has been studied in [210].

The idea of [145] has been generalized via an appropriate coordinate transformation to nonlinear systems, e.g., multiple-input-multiple-output (MIMO) nonminimum phase nonlinear systems in [142] and single-input-single-output (SISO) nonlinear systems with unmodeled dynamic in [146]. The semiglobal output regulation problem is addressed in [181] for nonlinear systems with unknown parameters in the exosystem using adaptive internal model. Global stabilization and disturbance suppression are achieved in [46] via output feedback with a new formulation of internal model, which is used to generate the feedforward compensation to a state variable, rather than the control compensation itself. The unknown parameters are dealt with in the control design by using the adaptive back-stepping technique. Output regulation with unknown linear exosystems is dealt with in [45] for uncertain nonlinear systems in the output feedback form, where the uncertain part is restricted in form of unknown parameters such that the adaptive technique may be designed to identify the unknown parameter. In [208], the uncertain nonlinear systems described in [45] are extended to some bounded function cases, such as polynomial functions.

It is noted that the stabilizing controller in the output regulation approaches is embedded with the internal model. An alternative disturbance rejection approach is to provide a separate or decoupling procedure for estimation and rejection, respectively.

One kind of valuable approaches applies adaptive techniques to deal with this type of disturbances. For linear systems, two adaptive algorithms are proposed in [12], where the first one is obtained by combining a frequency estimation algorithm together with an adaptive algorithm for the cancelation of disturbances of known frequency, and the second algorithm considers the frequency, phase, and magnitude of the disturbance in an integrated fashion. The algorithm was further extended to handle multiple sinusoidal components in [97]. An adaptive disturbance rejection scheme for nonlinear systems has also been provided in [158] by feedforward compensation combined with adaptive identification techniques. Furthermore, the adaptive servo-compensation of unknown disturbances was presented in [159]. In [147], the adaptive regulation problem in linear MIMO systems subject to unknown sinusoidal exogenous inputs is dealt with. The design of an adaptive regulator is performed within a set of Q-parameterized stabilizing controllers. It is noted that it is still an open problem to apply the adaptive technique for MIMO nonlinear systems with uncertainties.

As a class of strong-coupled nonlinear systems, the robotic system models suffer unmodeled dynamics and have nonlinear terms inherent in a mass matrix. Although the output regulation and adaptive compensation approaches have shown their theoretical and application significance, it is a challenging task to provide a feasible anti-disturbance control frame for nonlinear robotic systems with both harmonic disturbances and uncertainties.

The disturbance observer based control (DOBC), which originated from frequency domain methods for linear systems, has been widely used in many types of applications ([36, 87, 182, 193]). In many cases the DOBC and notable output regulation problem are fundamentally related (see survey paper [87] and references therein), where the disturbances are viewed as outputs of exosystems and may be rejected in the feedforward loop. Similarly to the adaptive algorithms, the heavy computation involved partial differential equations (PDEs) in output regulation can be avoided and the analysis can be provided hierarchically for disturbance estimation and compensation, respectively.

In [28], a nonlinear disturbance observer is constructed to estimate constant disturbances for one class of robotic systems. In [32], the sinusoidal disturbance acting on a single link of a robotic manipulator is estimated. In [169], adaptive frequency estimation technique is incorporated into the traditional disturbance observer method. For the uncertain nonlinearity case, a robust DOBC approach is firstly developed in [85] to enhance the robust performance for a class of MIMO nonlinear systems. When the plants have multiple disturbances and unmodeled dynamics, it has been shown that the composite hierarchical anti-disturbance control (CHADC) law can achieve the simultaneous disturbance attenuation and rejection performance for multiple disturbances ([93, 202]). Alternatively, Back and Shim proposed an add-on type inner loop output feedback controller for a class of SISO nonlinear system, such that the closed-loop system can be transformed into the standard singular perturbation form via a nonlinear disturbance observer [4]. However, up to now, most DOBC approaches for nonlinear systems only consider the disturbances with known exosystem models. As shown in [85], few DOBC results have been provided for MIMO nonlinear systems subject to disturbances with unknown frequencies and phases.

Our main purpose is to derive a generalized DOBC methodology for a class of nonlinear robotic systems. We provide a novel approach to observe the disturbance by constructing a two-step nonlinear disturbance observer (TSNLDO). Furthermore, even where an un-modeled dynamic is present, the satisfactory tracking behavior can be guaranteed through integrating the disturbance observers with conventional feedback control laws.

9.2   FORMULATION OF THE PROBLEM

Similarly to [132], the nonlinear robotic system considered in this chapter is described as:

$$M(q)\ddot{q} + h(\dot{q},q) = \tau + d \qquad (9.1)$$

where q is the $n \times 1$ dimensional displacement vector, and is supposed to be measurable together with $\dot{q}$, $M(q)$ is the $n \times n$ symmetric bounded positive definite nonlinear mass (or inertia) matrix, $h(\dot{q},q)$ is the $n \times 1$ vector of nonlinear terms, τ and $d = [d_1, d_2, \cdots, d_n]^T$ are $n \times 1$ vector of control input and disturbance with compatible dimensions.

$d_i(i = 1, 2 \ldots n)$ in (9.1) is formulated as:

$$d_i = \Phi_i sin(\omega_i t + \varphi_i), \qquad (9.2)$$

where $\Phi_i, \omega_i, \varphi_i$ are unknown. The harmonic disturbance in the control input path is a vector of sinusoidal signals with unknown amplitude, phase and frequency that can be formulated by the following exogenous system.

$$\begin{cases} \dot{w} = \Gamma w \\ d = V w \end{cases} \qquad (9.3)$$

where $w \in R^{2n}, \Gamma \in R^{2n \times 2n}$ has all its eigenvalues on the imaginary axis, and $V \in R^{n \times 2n}$ are proper unknown matrices. For simplicity, denote

$$\Gamma = \begin{bmatrix} \Gamma_1 & 0 & 0 \\ 0 & \ddots & 0 \\ 0 & 0 & \Gamma_n \end{bmatrix}, V = \begin{bmatrix} V_1 & 0 & 0 \\ 0 & \ddots & 0 \\ 0 & 0 & V_n \end{bmatrix}$$

and without loss of generality, suppose $\begin{bmatrix} \Gamma_i, & V_i \end{bmatrix}$ $(i = 1, 2, \ldots, n)$ has observable canonical form described as follows:

$$\Gamma_i = \begin{bmatrix} 0 & 1 \\ -\omega_i^2 & 0 \end{bmatrix}, V_i = \begin{bmatrix} 1 & 0 \end{bmatrix} \qquad (9.4)$$

The unknown parameters of frequency, amplitude and phase have relations with $\omega_i(i = 1 \ldots n)$ and the initial condition of Equation (9.3).

In this chapter we address the problem of rejecting single sinusoidal disturbances of unknown parameters for robotic systems (9.1). Also, we consider the uncertainty in $h(\dot{q}, q)$ described by

$$h(\dot{q}, q) = h_1(\dot{q}, q) - \Delta h(\dot{q}, q)$$

where $h_1(\dot{q}, q)$ and $\Delta h(\dot{q}, q)$ represent known and uncertainty parts in $h(\dot{q}, q)$, respectively. For any pair of $\begin{bmatrix} q \\ \dot{q} \end{bmatrix}$ and $\begin{bmatrix} \bar{q} \\ \dot{\bar{q}} \end{bmatrix}$, the nonlinear uncertain term is supposed to satisfy

$$\begin{cases} \Delta h(0,0) = 0 \\ \left\| \begin{bmatrix} \Delta h(\dot{q}, q) - \Delta h(\dot{\bar{q}}, \bar{q}) \\ M^{-1}(q)(\Delta h(\dot{q}, q) - \Delta h(\dot{\bar{q}}, \bar{q})) \end{bmatrix} \right\| \leq \\ \qquad \left\| \begin{bmatrix} U_{11} & 0 \\ 0 & U_{12} \end{bmatrix} \left(\begin{bmatrix} q \\ \dot{q} \end{bmatrix} - \begin{bmatrix} \bar{q} \\ \dot{\bar{q}} \end{bmatrix} \right) \right\| \end{cases} \qquad (9.5)$$

where U_{11} and U_{12} are given constant weighting matrix.

Thus, system (9.1) can be formulated as

$$M(q)\ddot{q} + h_1(\dot{q}, q) - \Delta h(\dot{q}, q) = \tau + d \qquad (9.6)$$

In general, for such robotic systems with nonlinear uncertain terms in a mass (or inertia) matrix, the output regulation theory is difficult to apply directly. In addition, up to now neither the present adaptive compensation approaches nor the DOBC approaches have been devoted to such MIMO nonlinear systems with unknown frequency harmonic disturbances and unmodeled uncertainty. In [28, 32, 85], nonlinear disturbance observers were constructed only for disturbances with known parameters. Also, it is challenging to analyze estimation stability in the presence of an additional unknown unmodeled dynamics.

In this chapter, we will establish a novel framework to realize the rejection of unknown disturbances for a class of nonlinear robotic systems. In the following, we construct a reduced-order observer for $d(t)$ with unknown parameters first, and then design a composite controller by combining the disturbance observer with a feedback controller. The stability of the resulting composite system can also be guaranteed when a system contains unmodeled dynamics.

9.3 TWO-STEP NONLINEAR DISTURBANCE OBSERVER

In most conventional DOBC strategies, the disturbance observer is constructed according to the disturbance dynamic model. A question arises regarding how the merits of DOBC can be inherited for the sinusoidal disturbances rejection problem with unknown frequencies and amplitudes. In this chapter, the following property of sinusoidal disturbances is required to construct a disturbance observer, which has been introduced in [158, 159] for SISO systems.

Lemma 9.1

Give an exosystem described by

$$\begin{cases} \dot{w}_0 &= \Gamma_0 w_0 \\ d_0 &= V_0 w_0 \end{cases} \tag{9.7}$$

where $w_0 \in R^p$ and $\Gamma_0 \in R^{p \times p}$, $[\Gamma_0, V_0]$ is observable, Γ_0 has all its eigenvalues on the imaginary axis and $V_0 = [v_{01}, v_{02}, \ldots, v_{0p}]$ is a constant vector. Then for any $p \times p$ Hurwitz matrix G_0, there exists a unique constant vector $\theta_0 \in R^p$ such that the signal d_0 can be presented in the form of

$$d_0 = \theta_0^T \xi_0 + \theta_0^T \delta_0,$$

where $\xi_0 \in R^p$, the dynamic system satisfies

$$\dot{\xi}_0 = G_0 \xi_0 + b_0 d_0,$$

and the pair (G_0, b_0) is controllable. ∎

The exponentially decaying vector δ_0 obeys

$$\dot{\delta}_0 = G_0 \delta_0$$

with $\delta_0(0) = M_0 w_0(0) - \xi_0(0)$.

The real value of θ_0 can be described by $\theta_0^T = V_0 M_0^{-1}$, where matrix M_0 is a solution of the Sylvester matrix equation

$$M_0 \Gamma_0 - G_0 M_0 = b_0 V_0. \tag{9.8}$$

This Lemma reduces the uncertainty of the harmonic signal d_0 to the parametric uncertainty of a constant vector θ_0. We will develop Lemma 9.1 to more general cases and generalize the disturbance estimation framework through construction of a two-step disturbance observer. The first-step observer gives the relation between the nonlinear functions ξ, δ and unknown constant vector θ with disturbance vector d. With the formulated disturbance, the task of the second-step observer is to estimate the unknown constant vector θ related to disturbance. The signatures ξ, δ and θ will be introduced in this section.

9.3.1 THE FIRST-STEP DISTURBANCE OBSERVER

Construct the following MIMO affine system

$$\begin{cases} \xi &= v + \psi \\ \dot{v} &= G(v + \psi) - L\frac{\partial M(q)}{\partial t}\dot{q} + Lh_1(\dot{q}, q) - L\tau \\ \psi &= LM(q)\dot{q} \end{cases} \tag{9.9}$$

where vectors ξ, ν, ψ have compatible dimensions and ξ is described as

$$\xi = \begin{bmatrix} \xi_1 \\ \xi_2 \\ \vdots \\ \xi_n \end{bmatrix}, \xi_i = \begin{bmatrix} \xi_{i1} \\ \xi_{i2} \end{bmatrix},$$

$G_i \in R^{2 \times 2}(i = 1 \ldots n)$ is Hurwitz, (G_i, L_i) is controllable, and G and L satisfy

$$G = \begin{bmatrix} G_1 & 0 & 0 \\ 0 & \ddots & 0 \\ 0 & 0 & G_n \end{bmatrix}, L = \begin{bmatrix} L_1 & 0 & 0 \\ 0 & \ddots & 0 \\ 0 & 0 & L_n \end{bmatrix}, \tag{9.10}$$

$G_i \in R^{2 \times 2}(i = 1 \ldots n)$ is a selectable parameter matrix to be constructed together with $L_i \in R^{2 \times 1}(i = 1 \ldots n)$.

Lemma 9.2

If there exists vector-functions ξ, ν and ψ satisfying (9.9), then the disturbance d in (9.3) can be presented in the form of

$$d = \Xi\theta + \Theta\delta \tag{9.11}$$

where

$$\Xi = \begin{bmatrix} \xi_1^T & 0 & 0 \\ 0 & \ddots & 0 \\ 0 & 0 & \xi_n^T \end{bmatrix}, \Theta = \begin{bmatrix} \theta_1^T & 0 & 0 \\ 0 & \ddots & 0 \\ 0 & 0 & \theta_n^T \end{bmatrix},$$

$$\theta = \begin{bmatrix} \theta_1 \\ \vdots \\ \theta_n \end{bmatrix}, \theta_i = \begin{bmatrix} \theta_{i1} & \theta_{i2} \end{bmatrix}^T,$$

$\theta_i(i = 1, 2 \ldots n)$ is an unknown constant vector to be determined and δ obeys the equation

$$\dot{\delta} = G\delta - L\Delta h(\dot{q}, q) \tag{9.12}$$

∎

Proof. From (9.9), it can be verified that

$$\dot{\xi} = G\xi + L(d + \Delta h(\dot{q}, q)). \tag{9.13}$$

Suppose that there exists vector $\hat{\xi} \in R^{2n \times 1}$ satisfying

$$\dot{\hat{\xi}} = G\hat{\xi} + Ld, \tag{9.14}$$

where

$$\hat{\xi} = \begin{bmatrix} \hat{\xi}_1 \\ \hat{\xi}_2 \\ \vdots \\ \hat{\xi}_n \end{bmatrix}, \hat{\xi}_i = \begin{bmatrix} \hat{\xi}_{i1} \\ \hat{\xi}_{i2} \end{bmatrix}$$

Based on Lemma 9.1, we know that θ or Θ as in (9.11) and $\hat{\delta} \in R^{2n \times 1}$ satisfying

$$\dot{\hat{\delta}} = G\hat{\delta}, \tag{9.15}$$

such that d may be expressed as

$$\begin{aligned} d &= \hat{\Xi}\theta + \Theta\hat{\delta} = (\hat{\Xi} - \Xi + \Xi)\theta + \Theta\hat{\delta} \\ &= \Xi\theta + \Theta\hat{\delta} + \Theta(\hat{\xi} - \xi) \end{aligned} \tag{9.16}$$

where

$$\hat{\Xi} = \begin{bmatrix} \hat{\xi}_1^T & 0 & 0 \\ 0 & \ddots & 0 \\ 0 & 0 & \hat{\xi}_n^T \end{bmatrix}$$

Based on (9.13), (9.14) and (9.15) it follows that

$$\dot{\hat{\xi}} - \dot{\xi} = G(\hat{\xi} - \xi) - L\Delta h(\dot{q}, q),$$

$$\dot{\hat{\xi}} - \dot{\xi} + \dot{\hat{\delta}} = G(\hat{\xi} - \xi + \hat{\delta}) - L\Delta h(\dot{q}, q).$$

Thus, together with Equation (9.16), it can be seen that

$$d = \Xi\theta + \Theta\delta$$

where $\delta = \hat{\xi} - \xi + \hat{\delta}$, and further

$$\dot{\delta} = G\delta - L\Delta h(\dot{q}, q).$$

This completes the proof.   □

9.3.2   THE SECOND-STEP DISTURBANCE OBSERVER

According to Lemma 9.2, Equation (9.1) can be rewritten as

$$M(q)\ddot{q} + h_1(\dot{q}, q) = \tau + \Xi\theta + \Theta\delta + \Delta h(\dot{q}, q), \tag{9.17}$$

and the major work of disturbance estimation remaining is how to identify the unknown constant vector θ. As there exist unmodeled dynamics $\Delta h(\dot{q}, q)$ in (9.17), many elegant adaptive methods are unfeasible to be implanted directly as in ([120, 159]). In the following, the unknown constant vector θ will be estimated by a nonlinear observer instead of adaptive algorithms. It will be seen that a fast convergence property can be achieved. To illustrate the procedure clearly, we examine the estimation of a single sinusoidal disturbance firstly. Suppose Γ_0 in (9.7) satisfies

$$\Gamma_0 = \begin{bmatrix} 0 & 1 \\ -\omega_0^2 & 0 \end{bmatrix}, V_0 = \begin{bmatrix} 1 & 0 \end{bmatrix}, \tag{9.18}$$

The 2×2 constant matrix Γ_0 depends on ω_0 which represents the unknown frequency of the harmonic signal d_0.

Theorem 9.1

Consider system $\dot{\xi}_0 = G_0\xi_0 + b_0 d_0$, where single sinusoidal disturbance d_0 is described as in (9.7), and suppose Γ_0 and V_0 satisfy (9.18).

If G_0 and b_0 are selected in form of

$$G_0 = \begin{bmatrix} 0 & 1 \\ -g_{01} & -g_{02} \end{bmatrix}, b_0 = \begin{bmatrix} 0 \\ 1 \end{bmatrix},$$

where g_{01} and g_{02} are constant values and guarantee G_0 is Hurwitz, then the harmonic signal d_0 can be described by

$$d_0 = [\theta_{01} \quad g_{02}]\xi_0 + [\theta_{01} \quad g_{02}]\delta_0, \tag{9.19}$$

where θ_{01} is an unknown constant value and the meaning of δ_0 is the same as in Lemma 9.1.   ∎

Proof. Applying Γ_0, G_0, V_0 and b_0 into the Sylvester matrix Equation (9.8), the unique invertible matrix M_0 is calculated as

$$M_0^{-1} = \begin{bmatrix} g_{01} - \omega_0^2 & g_{02} \\ -\omega_0^2 g_{02} & -\omega_0^2 + g_{01} \end{bmatrix}, \tag{9.20}$$

according to $\theta_0^T = V_0 M_0^{-1}$ as shown in Lemma 9.1, the conclusion can be obtained that $\theta_0 = [g_{01} - \omega_0^2 \quad g_{02}]^T$. Thus according to Lemma 9.1, (9.19) can be obtained easily. This completes the proof.   □

Based on Theorem 9.1, it can be seen that if $G_i(i = 1 \ldots n)$ and $L_i(i = 1 \ldots n)$ can be selected as

$$G_i = \begin{bmatrix} 0 & 1 \\ -g_{i1} & -g_{i2} \end{bmatrix}, L_i = \begin{bmatrix} 0 \\ 1 \end{bmatrix} \tag{9.21}$$

respectively, $\theta_{i2} = g_{i2}(i = 1 \ldots n)$ holds. Thus (9.11) can be rewritten as

$$d = \Xi_1 \bar{\theta}_1 + \Xi_2 \bar{g}_2 + \Theta \delta \tag{9.22}$$

where

$$\Xi_1 = \begin{bmatrix} \xi_{11} & 0 & 0 \\ 0 & \ddots & 0 \\ 0 & 0 & \xi_{n1} \end{bmatrix}, \Xi_2 = \begin{bmatrix} \xi_{12} & 0 & 0 \\ 0 & \ddots & 0 \\ 0 & 0 & \xi_{n2} \end{bmatrix}$$

$$\bar{g}_2 = \begin{bmatrix} g_{12} \\ \vdots \\ g_{n2} \end{bmatrix}, \bar{\theta}_1 = \begin{bmatrix} \theta_{11} \\ \vdots \\ \theta_{n1} \end{bmatrix}. \tag{9.23}$$

Thus, the substitution of (9.22) into (9.13) yields

$$\dot{\xi} = G\xi + L(\Xi_1 \bar{\theta}_1 + \Xi_2 \bar{g}_2 + \Theta \delta + \Delta h(\dot{q}, q)), \tag{9.24}$$

combining with (9.10), (9.21) and (9.24) we have

$$\begin{cases} \dot{\bar{\xi}}_1 = \bar{\xi}_2 \\ \dot{\bar{\xi}}_2 = -\Xi_1 \bar{g}_1 + \Xi_1 \bar{\theta}_1 + \Theta \delta + \Delta h(\dot{q}, q) \end{cases} \tag{9.25}$$

where

$$\bar{g}_1 = \begin{bmatrix} g_{11} \\ \vdots \\ g_{n1} \end{bmatrix}, \bar{\xi}_1 = \begin{bmatrix} \xi_{11} \\ \vdots \\ \xi_{n1} \end{bmatrix}, \bar{\xi}_2 = \begin{bmatrix} \xi_{12} \\ \vdots \\ \xi_{n2} \end{bmatrix}$$

From above results it is shown that through special selection of G and L in (9.21), the estimation of θ can be reduced to estimation of constant vector $\bar{\theta}_1$ according to Theorem 1. With the first-step observer (9.9), the components $\xi_{i1}(i = 1, 2 \ldots n)$ in nonlinear matrix Ξ_1 can be obtained, which will help us design the second-step observer for $\bar{\theta}_1$ based on (9.25).

Construct

$$\begin{cases} \hat{\bar{\theta}}_1 = z + p \\ \dot{z} = -\alpha \Xi_1^2 \hat{\bar{\theta}}_1 + \alpha \Xi_1^2 \bar{g}_1 - \alpha \Xi_2 \bar{\xi}_2 \\ p = \alpha \Xi_1 \bar{\xi}_2 \end{cases} \tag{9.26}$$

where $\alpha > 0$ is a given constant value. From (9.25) and (9.26) we can get

$$\begin{aligned}
\dot{\hat{\theta}}_1 &= -\alpha \Xi_1^2 \hat{\theta}_1 + \alpha \Xi_1^2 \bar{g}_1 + \alpha \Xi_2 \bar{\xi}_2 - \alpha \Xi_2 \bar{\xi}_2 \\
&\quad + \alpha \Xi_1(-\Xi_1 \bar{g}_1 + \Xi_1 \bar{\theta}_1 + \Theta \delta + \Delta h(\dot{q}, q)) \\
&= \alpha \Xi_1^2 \bar{\theta}_1 + \alpha \Xi_1 \Theta \delta + \alpha \Xi_1 \Delta h(\dot{q}, q).
\end{aligned}$$

Differentiating estimation error $\tilde{\bar{\theta}}_1 = \bar{\theta}_1 - \hat{\bar{\theta}}_1$, we obtain the following error model

$$\dot{\tilde{\bar{\theta}}}_1 = -\alpha \Xi_1^2 \tilde{\bar{\theta}}_1 - \alpha \Xi_1 \Theta \delta - \alpha \Xi_1 \Delta h(\dot{q}, q). \tag{9.27}$$

As $\Delta h(\dot{q}, q)$ in system model (9.6) is considered, the dynamic property of (9.27) turns out to be very complicated. Whereas, an advantage of the proposed DOBC frame is that $\Delta h(\dot{q}, q)$ does not need to be identified as in traditional adaptive algorithms. Instead, $\Delta h(\dot{q}, q)$ and d will be attenuated and rejected simultaneously by combining the observer with a robust controller. This property will be shown in the next section.

9.4 DOBC WITH STABILITY ANALYSIS

In the DOBC scheme, τ in (9.1) can be constructed as

$$\tau = M(q)u + h_1(\dot{q}, q) - \hat{d} \tag{9.28}$$

where $u \in R^n$ is the feedback controller to be designed, $\hat{d}$ is the disturbance observer and satisfies

$$\hat{d} = \Xi_1 \hat{\bar{\theta}}_1 + \Xi_2 \bar{g}_2 \tag{9.29}$$

Substituting (9.28) into (9.1) yields

$$\ddot{q} = u + M^{-1}(q)\Delta h(\dot{q}, q) + M^{-1}(q)\Xi_1 \tilde{\bar{\theta}}_1 + M^{-1}(q)\Theta \delta$$

Thus, the dynamic system (9.1) can be rewritten as

$$\begin{aligned}
\dot{x} &= Ax + F_1 f_M(x(t),t) f_1(x(t),t) + B(u + f_M(x(t),t)\Xi_1 \tilde{\bar{\theta}}_1 \\
&\quad + f_M(x(t),t)\Theta \delta)
\end{aligned} \tag{9.30}$$

where

$$x = \begin{bmatrix} q \\ \dot{q} \end{bmatrix}, f_1(x(t),t) = \Delta h(\dot{q}, q), f_M(x(t),t) = M^{-1}(q),$$

$$A = \begin{bmatrix} 0 & I \\ 0 & 0 \end{bmatrix}, B = F_1 = \begin{bmatrix} 0 \\ I \end{bmatrix},$$

the feedback controller $u = Kx$ and K is conventional control gain for stabilization. For reason of simplicity we denote the corresponding compatible matrices

$$\Pi(x(t),\Theta) = \Pi_0 + \Pi_1(x(t),\Theta) \tag{9.31}$$

where

$$\Pi(x(t),\Theta) = \begin{bmatrix} f_M(x(t),t) \\ f_M(x(t),t)\Theta \\ \Theta \end{bmatrix},$$

$$\Pi_0(x(t),\Theta) = \begin{bmatrix} M_0 \\ M_{\Theta 0} \\ \Theta_0 \end{bmatrix},$$

$$\Pi_1(x(t),\Theta) = \begin{bmatrix} M_1(x(t),t) \\ M_{\Theta 1}(x(t),t) \\ \Theta_1 \end{bmatrix}.$$

Π_0 and $\Pi_1(x,\Theta)$ represent nominal constant component and uncertain part of $\Pi(x,\Theta)$, respectively. As we know, $f_M(x(t),t)$ is a bounded positive function matrix, and Θ is a constant matrix related to disturbances frequencies $\omega_i(i = 1,2\dots n)$ in (9.4). From Theorem 9.1, if the regions of $\omega_i(i = 1,2\dots n)$ are given we can also figure out the region of Θ. Then according to (9.5), (9.31) and property of $f_M(x,t)$, we can find corresponding weighting matrices $U_i(i = 1,2\dots 4)$ satisfying

$$\|M_{\Theta 1}(x,t)\delta\| \le \|U_3\delta\|,$$

$$\|M_1(x,t)\Xi_1\tilde{\bar{\theta}}_1\| \le \|U_2\Xi_1\tilde{\bar{\theta}}_1\|, \|\Theta_1\delta\| \le \|U_4\delta\|. \tag{9.32}$$

The composite system includes (9.12), (9.27) and (9.30) yields

$$\dot{\bar{x}} = \bar{A}\bar{x} + \bar{F}\bar{f}(\bar{x},t) \tag{9.33}$$

where

$$\bar{x} = \begin{bmatrix} x \\ \tilde{\bar{\theta}}_1 \\ \delta \end{bmatrix}, \bar{f}(\bar{x},t) = \begin{bmatrix} f_1(x(t),t) \\ f_M(x(t),t)f_1(x(t),t) \\ M_1(x(t),t)\Xi_1\tilde{\bar{\theta}}_1 \\ M_{\Theta 1}(x,t)\delta \\ \Theta_1\delta \end{bmatrix}$$

$$\bar{A} = \begin{bmatrix} A+BK & BM_0\Xi_1 & BM_{\Theta 0} \\ 0 & -\alpha\Xi_1^2 & -\alpha\Xi_1\Theta_0 \\ 0 & 0 & G \end{bmatrix} \tag{9.34}$$

$$\bar{F} = \begin{bmatrix} 0 & F_1 & B & B & 0 \\ -\alpha\Xi_1 & 0 & 0 & 0 & -\alpha\Xi_1 \\ -L & 0 & 0 & 0 & 0 \end{bmatrix}$$

Thus, according to (9.5) and (9.32) it can be derived that there exists weighting matrix U satisfying $\|\bar{f}(\bar{x},t)\| < \|U\bar{x}\|$, where

$$U = \begin{bmatrix} U_{11} & 0 & 0 \\ U_{12} & 0 & 0 \\ 0 & U_2\Xi_1 & 0 \\ 0 & 0 & U_3 \\ 0 & 0 & U_4 \end{bmatrix} \tag{9.35}$$

At this stage, our objective is to find K such that the closed-loop system (9.33) with $u = Kx$ is asymptotically stable. For the sake of simplifying descriptions, we denote

$$\begin{aligned}
N_1 &= P_1(A+BK) + (A+BK)^T P_1 + \frac{1}{\lambda^2}U_{11}^T U_{11} + \frac{1}{\lambda^2}U_{12}^T U_{12}, \\
N_2 &= \Xi_1^T \hat{N}_2 \Xi_1 \\
\hat{N}_2 &= -2R_2 I + \frac{1}{\lambda^2}U_2^T U_2 \\
N_3 &= Q_3 G + G^T Q_3 + \frac{1}{\lambda^2}U_4^T U_4 + \frac{1}{\lambda^2}U_3^T U_3
\end{aligned} \tag{9.36}$$

Theorem 9.2

For given $\lambda > 0$, if there exists $Q_1 > 0, Q_3 > 0$ and $R_1, R_2 \in R$ satisfying

$$\Omega = \begin{bmatrix} \Omega_1 & \Omega_2 & \Omega_3 \\ * & -I & 0 \\ * & 0 & -I \end{bmatrix} < 0 \tag{9.37}$$

where

$$\Omega_1 = \begin{bmatrix} \Omega_{11} & BM_0 & BM_{\Theta 0} \\ * & \Omega_{12} & -R_2\Theta_0 \\ ** & * & \Omega_{13} \end{bmatrix},$$

$$\Omega_{11} = sym((AQ_1 + BR_1)), \Omega_{12} = -2R_2 I, \Omega_{13} = sym(G^T Q_3)$$

$$\Omega_2 = \lambda \begin{bmatrix} 0 & F_1 & B & B & 0 \\ -R_2 I & 0 & 0 & 0 & -R_2 I \\ -Q_3 L & 0 & 0 & 0 & 0 \end{bmatrix}$$

$$\Omega_3 = \begin{bmatrix} \frac{1}{\lambda}Q_1U_{11}^T & \frac{1}{\lambda}Q_1U_{12}^T & 0 & 0 & 0 \\ 0 & 0 & \frac{1}{\lambda}U_2^T & 0 & 0 \\ 0 & 0 & 0 & \frac{1}{\lambda}U_4^T & \frac{1}{\lambda}U_3^T \end{bmatrix},$$

then for arbitrary given $Q_2 > 0$ with compatible dimensions and under DOBC law (9.28), the composite system (9.33) with $K = R_1Q_1^{-1}, \alpha = R_2Q_2$ is asymptotically stable. ∎

Proof. After letting $P_1 = Q_1^{-1}, P_2 = Q_2^{-1}$, denote

$$V(\bar{x},t) = \bar{x}^TP\bar{x} + \frac{1}{\lambda^2}\int_0^t (\|U\bar{x}(\varepsilon)\|^2 - \|\bar{f}(\bar{x},\varepsilon)\|^2)d\varepsilon$$

where

$$P = \begin{bmatrix} P_1 & 0 & 0 \\ 0 & P_2I & 0 \\ 0 & 0 & Q_3 \end{bmatrix}, \tag{9.38}$$

then

$$\dot{V}(\bar{x},t) = \bar{x}^T(P\bar{A} + \bar{A}^TP)\bar{x} + 2\bar{x}^TP\bar{F}\bar{f}(\bar{x},t) +$$

$$\frac{1}{\lambda^2}\|U\bar{x}\|^2 - \frac{1}{\lambda^2}\|\bar{f}(\bar{x},t)\|^2$$

$$= \begin{bmatrix} \bar{x} \\ \frac{\bar{f}(\bar{x})}{\lambda} \end{bmatrix}^T \Lambda \begin{bmatrix} \bar{x} \\ \frac{\bar{f}(\bar{x})}{\lambda} \end{bmatrix} \tag{9.39}$$

where

$$\Lambda = \begin{bmatrix} sym(P\bar{A}) + \frac{1}{\lambda^2}U^TU & \lambda P\bar{F} \\ * & -I \end{bmatrix} \tag{9.40}$$

Substituting (9.34), (9.35) and (9.38) into (9.40) yields

$$\Lambda = \begin{bmatrix} \bar{\Omega}_1 & \bar{\Omega}_2 \\ * & -I \end{bmatrix} \tag{9.41}$$

where

$$\bar{\Omega}_1 = \begin{bmatrix} N_1 & P_1BM_0\Xi_1 & P_1BM_{\Theta0} \\ * & N_2 & -R_2\Xi_1\Theta_0 \\ * & * & N_3 \end{bmatrix},$$

$$\bar{\Omega}_2 = \lambda \begin{bmatrix} 0 & P_1F_1 & P_1B & P_1B & 0 \\ -R_2\Xi_1 & 0 & 0 & 0 & -R_2\Xi_1 \\ -Q_3L & 0 & 0 & 0 & 0 \end{bmatrix}$$

Rewrite Λ as

$$\Lambda = \bar{\Lambda}^T \Lambda_1 \bar{\Lambda} \tag{9.42}$$

where

$$\bar{\Lambda} = \begin{bmatrix} I & 0 & 0 & 0 & 0 & 0 & 0 & 0 \\ 0 & \Xi_1 & 0 & 0 & 0 & 0 & 0 & 0 \\ 0 & 0 & I & 0 & 0 & 0 & 0 & 0 \\ 0 & 0 & 0 & I & 0 & 0 & 0 & 0 \\ 0 & 0 & 0 & 0 & I & 0 & 0 & 0 \\ 0 & 0 & 0 & 0 & 0 & I & 0 & 0 \\ 0 & 0 & 0 & 0 & 0 & 0 & I & 0 \\ 0 & 0 & 0 & 0 & 0 & 0 & 0 & I \end{bmatrix}, \Lambda_1 = \begin{bmatrix} \bar{\bar{\Omega}}_1 & \bar{\bar{\Omega}}_2 \\ * & I \end{bmatrix},$$

$$\bar{\bar{\Omega}}_1 = \begin{bmatrix} N_1 & P_1 BM_0 & P_1 BM_{\Theta 0} \\ * & \hat{N}_2 & -R_2 \Theta_0 \\ * & * & N_3 \end{bmatrix}$$

$$\bar{\bar{\Omega}}_2 = \lambda \begin{bmatrix} 0 & P_1 F_1 & P_1 B & P_1 B & 0 \\ -R_2 & 0 & 0 & 0 & -R_2 \\ -Q_3 L & 0 & 0 & 0 & 0 \end{bmatrix}$$

Pre-multiplied and post-multiplied simultaneously by diag $\{Q, I\}$, thus $\Lambda_1 < 0$ is equivalent to $\Lambda_2 < 0$, where

$$\Lambda_2 = \begin{bmatrix} Q\bar{\bar{\Omega}}_1 Q & Q\bar{\bar{\Omega}}_2 \\ * & I \end{bmatrix},$$

and

$$Q = \begin{bmatrix} Q_1 & 0 & 0 \\ 0 & I & 0 \\ 0 & 0 & I \end{bmatrix}$$

Based on Schur complement, it can be seen that $\Lambda_2 < 0$ is equivalent to $\Omega < 0$. The above conclusion means $\tilde{\bar{\theta}}_1$, x and δ are bounded, as Ξ_1^2 is not always larger than zero, the asymptotic stable conclusion cannot be achieved directly. But it is easy to get that $\ddot{V}(\bar{x}, t)$ is also bounded, thus, according to Barbalat lamma, $\tilde{\bar{\theta}}_1, \delta, x \to 0$ as $t \to +\infty$, then the proof is completed. $\qquad \square$

It is noticed that as Q_2 can be chosen arbitrarily, we can determine the value of α as desired, which influences the convergence property of $\tilde{\bar{\theta}}_1$ according to (9.27). The two-step observer based control design procedure can be summarized as follows:

1. Select weighting matrices G and L with form of (9.10) and (9.21). Apply G and L into Theorem 9.2, and solve (9.37) to obtain the feedback controller gain K and α.

2. Design the first-step observer for ξ according to (9.9) such that (9.13) is satisfied. Construct the second-step observer based on (9.26) to estimate unknown constant vector $\bar{\theta}_1$. Then the disturbance observer $\hat{d}$ can be derived as (9.29).

3. Apply $\hat{d}$ and feedback controller $u = Kx$ into (9.27), DOBC control torque τ can be realized.

9.5 SIMULATION

To show the efficiency of the proposed scheme, a two-link robotic manipulator is considered in this chapter. Up to now, few disturbance rejection schemes such as output regulation and DOBC theory have been introduced to deal with this class of nonlinear robotic systems. In this section, an adaptive strategy similar to [159] is adopted in this system to compare with our presented method.

The model of a two-link robotic manipulator can be represented by

$$M(q)(\ddot{q}) + C(\dot{q},q)\dot{q} + G(q) = \tau + d \qquad (9.43)$$

where

$$q = \begin{bmatrix} q_1 \\ q_2 \end{bmatrix}$$

is the 2×1 vector of joint angular positions, τ is a 2×1 vector of torques applied to the joints, $M(q)$ is the 2×2 symmetric bounded positive definite mass (or inertia) matrix, $\dot{q}$ is the 2×1 vector of joint angular velocity, $C(q,\dot{q})\dot{q}$ is the 2×1 vector of centrifugal and Coriolis terms, and $G(q)$ is the 2×1 gravity vector, d_0 is a disturbance torque or force vector. For the sake of simplicity, the mass matrices are given as follows, and can be readily extended to the more general case.

$$M(q) = \begin{bmatrix} p_1 + p_2 + 2p_2cos(q_2) & p_3 + p_2cos(q_2) \\ p_3 + p_2cos(q_2) & p_3 \end{bmatrix}.$$

Thus the $h(\dot{q},q)$ in (9.1) can be expressed as the vectors of centrifugal and Coriolis terms and gravity, i.e.,

$$h(q,\dot{q}) \quad = \quad C(q,\dot{q})\dot{q} + G(q)$$

Set $h_1(q,\dot{q})$ is the known part of $h(q,\dot{q})$ and

$$h_1(q,\dot{q}) = \begin{bmatrix} h_1 \\ h_2 \end{bmatrix} = \begin{bmatrix} -p_3(2\dot{q}_1\dot{q}_2 + \dot{q}_2^2)sin(q_2) \\ p_3\dot{q}_1^2 sin(q_2) \end{bmatrix},$$

where

$$p_1 = (m_1/4 + m_2)l_1^2 + I_1,$$

$$p_2 = m_2 l_1 l_2 / 2, p_3 = m_2 l_2^2 / 4 + I_2.$$

TABLE 9.1

The parameters in a two-link manipulator system.

Parameter	Symbol	Value
Mass of the i_{th} arm	$m_i(i=1,2)$	$8\ kg$
The i_{th} arm's moment inertia term	$I_i(i=1,2)$	$0.4\ kgm^2$
Length of the i_{th} arm	$l_i(i=1,2)$	$0.5\ m$

Source: Reference [215].

The significance of the parameters in (9.1) are given in Table 9.1 as in [215]. Exogenous system $d_i(i=1,2)$ shows disturbance caused by an actuator, expressed as nonlinear function $\Phi sin(\omega t + \varphi)$, where Φ, ω and φ are unknown as well as $w(0)$.

When DOBC law is applied in system (9.43), the corresponding parameter in (9.30) can be achieved at:

$$A = \begin{bmatrix} 0 & 0 & 1 & 0 \\ 0 & 0 & 0 & 1 \\ 0 & 0 & 0 & 0 \\ 0 & 0 & 0 & 0 \end{bmatrix},$$

$$B = \begin{bmatrix} 0 & 0 \\ 0 & 0 \\ 1 & 0 \\ 0 & 1 \end{bmatrix}, F_1 = \begin{bmatrix} 0 & 0 \\ 0 & 0 \\ 1 & 0 \\ 0 & 1 \end{bmatrix}$$

In the following, we will provide two comparisons with the DOBC and the present adaptive results. Firstly, we suppose

$$d = \begin{bmatrix} sin(4t) \\ cos(4t) \end{bmatrix}$$

and in this simulation, suppose

$$\|\Delta h(\dot{q}, q)\| \le \left\| \begin{bmatrix} 0.1 & 0 \\ 0 & 0.1 \end{bmatrix} \begin{bmatrix} \dot{q}_1 \\ \dot{q}_2 \end{bmatrix} \right\|$$

From Table 9.1, we can get the region of $M^{-1}(q)$, and then select nominal part of $M^{-1}(q)$ as

$$M_0 = \begin{bmatrix} 0.5 & -1 \\ -1 & 3 \end{bmatrix},$$

and together with definition of (9.31), $U_i(i = 1, 2 \ldots 4)$ in (9.32) is set as

$$U_{11} = \begin{bmatrix} 0 & 0 & 0.4 & 0 \\ 0 & 0 & 0 & 0.4 \end{bmatrix},$$

$$U_{12} = \begin{bmatrix} 0 & 0 & 2 & 0 \\ 0 & 0 & 0 & 2 \end{bmatrix}, U_2 = \begin{bmatrix} 0.5 & 0 \\ 0 & 0.5 \end{bmatrix},$$

$$U_3 = \begin{bmatrix} 4 & 0 & 0 & 0 \\ 0 & 3 & 0 & 0 \\ 0 & 0 & 10 & 0 \\ 0 & 0 & 0 & 7 \end{bmatrix},$$

$$U_4 = \begin{bmatrix} 15 & 0 & 0 & 0 \\ 0 & 0 & 15 & 0 \end{bmatrix}$$

G and L in (9.9) are selected as

$$G = \begin{bmatrix} 0 & 1 & 0 & 0 \\ -25 & -10 & 0 & 0 \\ 0 & 0 & 0 & 1 \\ 0 & 0 & -25 & -10 \end{bmatrix},$$

$$L = \begin{bmatrix} 0 & 0 \\ 1 & 0 \\ 0 & 0 \\ 0 & 1 \end{bmatrix}$$

Thus the nominal parts of Θ and $f_M(x(t), t)\Theta$ can be given as:

$$\Theta_0 = \begin{bmatrix} 30 & 10 & 0 & 0 \\ 0 & 0 & 30 & 10 \end{bmatrix},$$

$$M_{\Theta 0} = \begin{bmatrix} 13 & 5 & -25 & -10 \\ -25 & -10 & 80 & 32 \end{bmatrix}.$$

Based on Theorem 9.2 it can be solved that

$$K = \begin{bmatrix} -1.4223 & 0.1844 & -3.0930 & 0.3491 \\ 0.1431 & -2.2115 & 0.3453 & -2.5113 \end{bmatrix},$$

and $\alpha = 5000$. By the construction of the observers (9.9) and (9.26), we can get $\hat{\bar{\theta}}_1$ which is estimation value of $\bar{\theta}_1$. Then DOBC control torque τ (9.28) is derived with $u = Kx$ and $\hat{d}$ in (9.29).

Figures 9.1-9.3 demonstrate the system performance using the proposed two-step disturbance observer based scheme. The disturbance estimation error is less than 10^{-3} after 3 seconds (Figure 9.1). Correspondingly, the precise of control can be guaranteed by combining the disturbance observer with robust feedback controller (Figures 9.2-9.3).

When an adaptive compensation strategy is considered, we suppose the un-modeled dynamics are absent, i.e., $\Delta h(\dot{q},q) = 0$. Figures 9.4-9.6 show the disturbance estimation and control performances using an adaptive compensation approach. Theoretically, the system output can converge to zero as $t \to +\infty$. Whereas, from Figure 9.4 it can be seen that disturbances tracking error remains relatively large after 30 seconds with adaptive identification. As a result, the system output convergence property is unsatisfactory even when there is no unmodeled dynamics (Figures 9.5-9.7).

Secondly, we suppose d has two distinct frequency, i.e.,

$$d = \left[\begin{array}{c} sin(6t + \pi/4) \\ cos(8t + \pi/3) \end{array} \right]$$

we adopt the same parameter α and K in the simulation. It shows that the disturbance d can be estimated with similarly to the case where $\omega_{1,2} = 4$.

Thirdly, we consider the rejection of sinusoidal disturbances with unknown high frequencies. For the sake of simplicity, suppose $\omega_{1,2} = 100$ and $\Delta h(\dot{q},q) = 0$. From system (9.12) and (9.27) the disturbance estimation asymptotically stable property can be proven easily, and the convergence can be analyzed by selection of α in (9.30). Figures 9.8-9.10 demonstrate the satisfactory disturbance tracking performance and control precision by construction of the proposed two-step disturbance observer. However, an adaptive algorithm is unfeasible for estimating such high-frequency disturbances due to slow estimation convergence speed (Figure 9.11).

The simulation results show that although there are disturbances and modeling uncertainties in the system simultaneously, the sinusoidal disturbance rejection performance is improved and satisfactory system responses can be achieved using our presented method.

9.6 CONCLUSION

Anti-disturbance control for dynamic systems subject to disturbances and unmodeled dynamics is a fundamental problem in control theory. Up to this date, the internal model control, output regulation, adaptive compensation and DOBC approaches have focused on such problems and have been applied to many practical fields. However, it is still an ongoing problem for MIMO nonlinear robotic systems subject to unknown sinusoidal disturbances and unmodeled dynamics. This chapter gives a novel DOBC method for inherent MIMO nonlinear robotic systems subject to unknown sinusoidal disturbances and unmodeled dynamics. The proposed method systematically converts this problem to a kind of DOBC design problem by constructing a two-step disturbance observer. Both the nonlinear disturbance observer (NLDO) and controller are designed only by means of algebraic calculations or parameter selections.

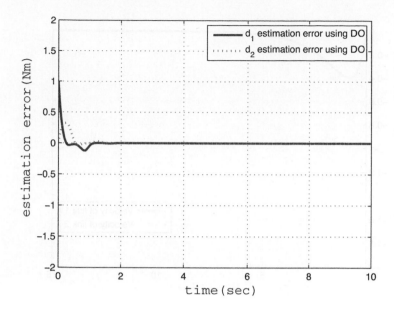

FIGURE 9.1   The estimation error of d_1 and d_2.

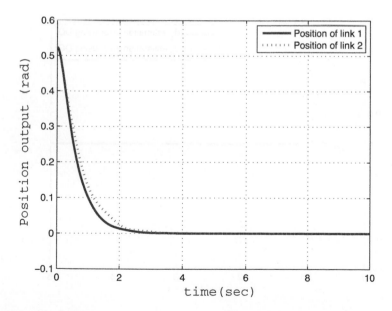

FIGURE 9.2   Tracking error of position using DOBC.

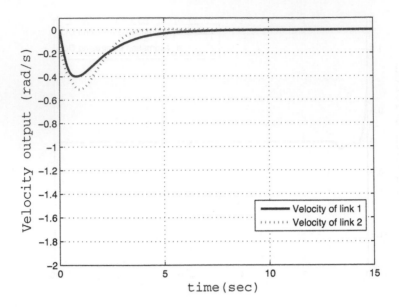

FIGURE 9.3   Tracking error of velocity using DOBC.

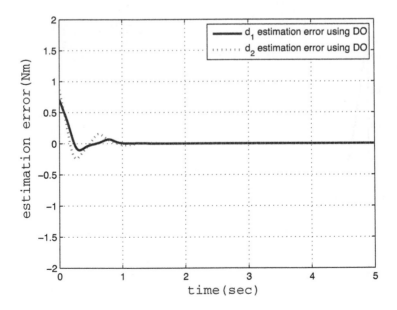

FIGURE 9.4   The estimation error of d_1 and $d_2(\omega_1 = 6, \omega_2 = 8)$.

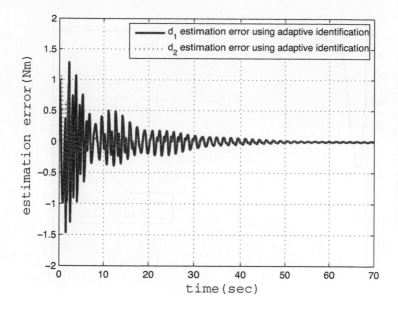

FIGURE 9.5   The adaptive identification error of d_1 and d_2.

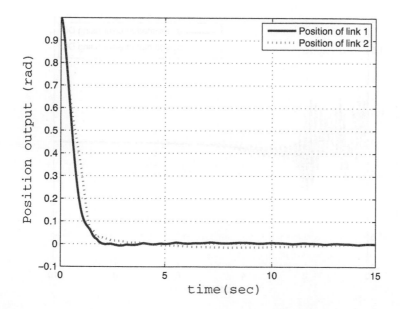

FIGURE 9.6   Tracking error of position using adaptive compensation.

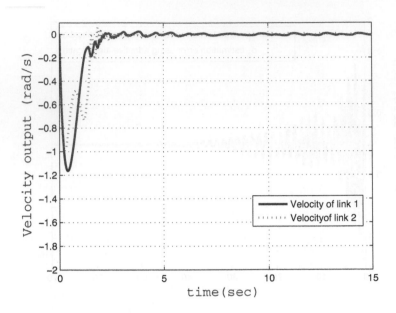

FIGURE 9.7   Tracking error of velocity using adaptive compensation.

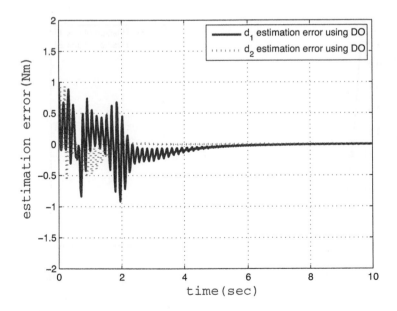

FIGURE 9.8   The estimation error of d_1 and d_2 with high frequency ($\omega_{1,2} = 100$).

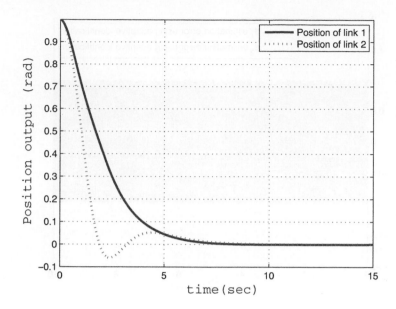

FIGURE 9.9   Tracking error of position using DOBC ($\omega_{1,2} = 100$).

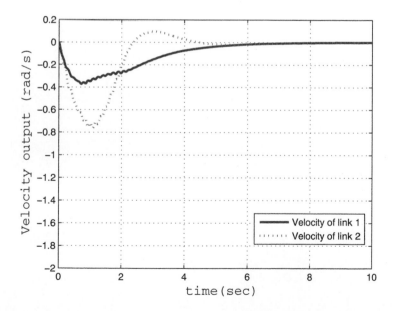

FIGURE 9.10   Tracking error of velocity using DOBC ($\omega_{1,2} = 100$).

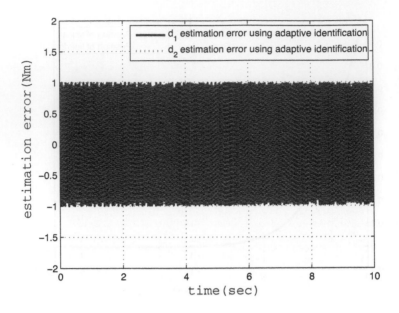

FIGURE 9.11   The adaptive identification error of d_1 and d_2 ($\omega_{1,2} = 100$).

10 Anti-Disturbance PD Attitude Control and Stabilization of Flexible Spacecraft

10.1 INTRODUCTION

High-precision attitude control for flexible spacecraft has been a difficult problem especially in communication, navigation, remote sensing, and other space-related missions. This is because modern spacecraft often employ large, deployed and light damping structures (such as solar paddles and antenna reflectors) to provide sufficient power supply and reduce launch costs. Consequently, the complex space structure may lead to decreased rigidity and low-frequency elastic modes. The dynamic model of a flexible spacecraft usually includes interaction between the rigid and elastic modes [149, 154]. The unwanted excitation of the flexible modes during the control of the rigid body attitude, together with other external disturbances, measurement and actuator error, and unmodeled dynamics, may cause degradation of the performance of attitude control systems (ACSs). Thus, the control scheme must provide not only adequate stiffness and damping to the rigid body modes, but also actively damp or reject the flexible modes. The desired control scheme must be robust enough to overcome the model uncertainty and unmodeled nonlinearity, various disturbances from the environment and structural vibrations of flexible appendages. In conclusion, it is a challenging control problem to achieve the requested pointing precision for flexible spacecraft.

Most present ACSs of spacecraft are based on proportional integral derivative (PID) or PD laws for their simplicity and reliability. When model uncertainties and coupling vibrations exist or external disturbance varies, it is difficult for PID/PD controllers to get satisfactory performance for flexible spacecraft. Since the 1990s, instead of the conventional PID/PD control, optimal control of flexible spacecraft has been studied for vibration suppression problems in [9, 14]. It is noted that robustness cannot be guaranteed for optimal control in the presence of model uncertainties and nonlinearity. From then on, "robust"control schemes have been designed for the attitude control and vibration suppression problem. Sliding mode control (SMC), known as an efficient and simple control strategy to systems with strong nonlinearity and model uncertainty, has been effectively applied to ACS design [37, 42, 187]. SMC is combined with active vibration control by piezoelectric actuator, but the chattering phenomenon caused by SMC has limited its practical applications. In [74, 75], the influence of flexibility on rigid motion, the presence of disturbances acting on the structure, and parameter variations have been considered in robust controller design. In [104, 105], a series of variable structure control schemes have been provided systematically for vibration suppression problems. H_∞ control has been used in ACS design in [5, 18] where external disturbance and model uncertainty are considered. An H_∞ multi-objective controller based on the linear matrix inequality (LMI) framework, is designed for flexible spacecraft in [25]. It is shown that robust controllers may lead to large conservativeness for the active vibration control problem, especially in case of fast dynamic disturbance and high-frequency perturbations.

It has been shown that DOBC may have less conservativeness for many types of disturbances and is easy to integrate with other conventional feedback controllers such as PD, H_∞, and variable structure controllers [85, 201]. In this chapter, we will design a composite controller based on DOBC and PD control schemes for flexible spacecraft, where DOBC can compensate for the effect of vibration from flexible appendages, and a PD controller can control the attitude of the spacecraft. Different from [85, 201], the disturbance is not confined to a constant, harmonic signal or a norm-bounded variable. With the proposed method, some new composite anti-disturbance methods (by combining DOBC with other feedback controllers) can be provided for ACS of spacecraft. Simulations for a flexible spacecraft shows that the performance of ACS can be guaranteed by the proposed method.

10.2   PROBLEM FORMULATION

To simplify the problem, only single-axis rotation is considered. We can obtain the single-axis model derived from the nonlinear attitude dynamics of the flexible spacecraft. It is assumed that this model includes one rigid body and one flexible appendage (see Figure 10.1) and the relative elastic spacecraft model is described as follows

$$J\ddot{\theta} + F\ddot{\eta} = u + w, \tag{10.1}$$

$$\ddot{\eta} + 2\xi\omega\dot{\eta} + \omega^2\eta + F^T\ddot{\theta} = 0, \tag{10.2}$$

where θ is the attitude angle, J is the moment of inertia of the spacecraft, F is the rigid-elastic coupling matrix, u is the control torque, w represents the merged disturbance torque including the space environmental torques, unmodeled uncertainties and noise from sensors and actuators, η is the flexible modal coordinate, ξ is the damping ratio, and ω is the modal frequency. Since vibration energy is concentrated in low-frequency modes in a flexible structure, its reduced-order model can be obtained by modal truncation. In this chapter, only the first two bending modes are taken into account.

Combining (10.1) with (10.2), we can get

$$(J - FF^T)\ddot{\theta} = F(2\xi\omega\dot{\eta} + \omega^2\eta) + u + w. \tag{10.3}$$

To (10.3), we consider $F(2\xi\omega\dot{\eta} + \omega^2\eta)$ as the disturbance due to elastic vibration of the flexible appendages. Denote $x(t) = \left[\theta(t), \dot{\theta}(t)\right]^T$, then (10.3) can be transformed into the following form

$$\dot{x}(t) = Ax(t) + B_u u(t) + B_f d_0(t) + B_d d_1(t), \tag{10.4}$$

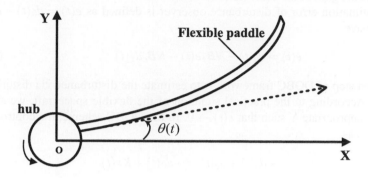

FIGURE 10.1 Spacecraft with flexible appendages.

where the coefficient matrices are denoted by

$$A = \begin{bmatrix} 0 & 1 \\ 0 & 0 \end{bmatrix}, B_u = B_f = B_d = \begin{bmatrix} 0 \\ (J - FF^T)^{-1} \end{bmatrix},$$

and

$$d_0(t) = F(2\xi\omega\dot{\eta} + \omega^2\eta)$$

is the disturbance from the flexible appendages, $d_1(t)$ is the "equivalent disturbance" including the space environmental disturbances, unmodeled uncertainties and noise from sensors and actuators, and $u(t)$ is the control input.

Obviously, the disturbance considered in this chapter generalizes the disturbance types studied in previous works. Suppose $\|\dot{d}_0(t)\| \leq W_0$ and $\|d_1(t)\| \leq W_1$, where W_0 and W_1 are known positive constants.

As the PD controller has been widely employed in many practical systems, we consider the classical PD controller described by

$$u_c(t) = K_P\theta(t) + K_D\dot{\theta}(t)$$

Denote $K = \begin{bmatrix} K_P & K_D \end{bmatrix}$, where K is the control gain to be determined.

10.3   COMPOSITE CONTROLLER DESIGN

10.3.1   DISTURBANCE OBSERVER DESIGN

According to system (10.4), we formulate the disturbance observer as

$$\begin{cases} \dot{\tau}(t) & = & -NB_f(\tau + Nx(t)) - N(Ax(t) + B_u u(t)) \\ \hat{d}_0(t) & = & \tau + Nx(t) \end{cases} \tag{10.5}$$

where N is the gain of the observer to be designed.

The estimation error of disturbance observer is defined as $e(t) = d_0(t) - \hat{d}_0(t)$. Then we have

$$\dot{e}(t) = \dot{d}_0(t) - NB_f e(t) - NB_d d_1(t) \tag{10.6}$$

The first step of DOBC framework is to estimate the disturbance via disturbance observer. According to the practical situation of the flexible spacecraft, we should design an appropriate N such that $e(t) \to 0$. In the DOBC scheme, the controller is constructed as

$$u(t) = -\hat{d}_0(t) + u_c(t) = -\hat{d}_0(t) + Kx(t)$$

The DOBC scheme can be described by Figure 10.2 (where $\hat{d}_0(t)$ is the estimation of $d_0(t)$). From Figure 10.2, it can be seen that the composite hierarchical controller consists of two parts, the inner loop is the disturbance observer and feedforward compensation, and the outside loop is the PD attitude controller. Thus, the composite controller can effectively control the spacecraft attitude and attenuate disturbances. Vibration caused by the flexible appendages is observed and compensated for, and the attitude of the spacecraft is controlled by PD controller.

Substituting $u(t)$ to (10.4) and (10.6), we can get the augmented closed-loop system

$$\begin{bmatrix} \dot{x}(t) \\ \dot{e}(t) \end{bmatrix} = \begin{bmatrix} A + B_u K & B_f \\ 0 & -NB_f \end{bmatrix} \begin{bmatrix} x(t) \\ e(t) \end{bmatrix} + \begin{bmatrix} 0 & B_d \\ I & -NB_d \end{bmatrix} \begin{bmatrix} \dot{d}_0(t) \\ d_1(t) \end{bmatrix} \tag{10.7}$$

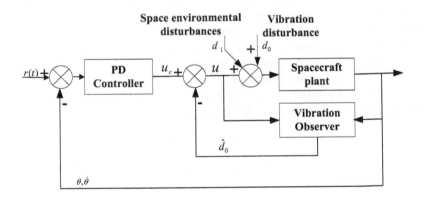

FIGURE 10.2   Block diagram of composite attitude controller.

10.3.2   STABILITY OF THE COMPOSITE SYSTEMS

In this section, we will consider the uniform ultimate boundedness (UUB) of the augmented system (10.7). We will give an LMI based design method to compute the controller gain and the observer gain, simultaneously.

Theorem 10.1

To augmented system (10.7), for $\alpha_1 > 0$, $\beta_1 > 0$ and $\beta_2 > 0$, if there exist matrices $Q_1 > 0$, $P_2 > 0$, R_1, and Q_2 satisfying

$$\Theta = \begin{bmatrix} \Theta_{11} & B_f & 0 & 0 \\ * & \Theta_{22} & Q_2 & P_2 \\ * & * & -\beta_1^{-2}I & 0 \\ * & * & * & -\beta_2^{-2}I \end{bmatrix} < 0 \tag{10.8}$$

where

$$\Theta_{11} = sym(AQ_1 + B_uR_1) + \alpha_1^2 I,$$

$$\Theta_{22} = -sym(Q_2B_f),$$

then the composite system (10.7) is uniformly ultimately bounded, the observer gain and the controller gain can be computed via $N = P_2^{-1}Q_2$ and $K = R_1Q_1^{-1}$. ∎

Proof. Denote

$$\begin{aligned} V(x,e,t) &= \begin{bmatrix} x^T & e^T \end{bmatrix} \begin{bmatrix} P_1 & 0 \\ 0 & P_2 \end{bmatrix} \begin{bmatrix} x \\ e \end{bmatrix} \\ &= x^T P_1 x + e^T P_2 e \\ &= V_1(x,t) + V_2(e,t) \end{aligned}$$

Computing the derivative of $V_1(x,t)$ along the trajectories of (10.7), we can obtain

$$\begin{aligned} \dot{V}_1(x,t) &= x^T(sym(P_1A + P_1B_uK))x + 2x^T P_1 B_f e \\ &\quad + 2x^T P_1 B_d d_1 \\ &\leq x^T(sym(P_1A + P_1B_uK) \\ &\quad + \alpha_1^2 P_1 P_1^T)x + \frac{1}{\alpha_1^2}\|B_d\|^2\|d_1\|^2 + 2x^T P_1 B_f e \end{aligned}$$

Similarly, we can compute the derivative of $V_2(e,t)$ in the following

$$
\begin{aligned}
\dot{V}_2(e,t) &= e^T P_2 \dot{e} + \dot{e}^T P_2 e \\
&= sym(e^T P_2 \dot{d}_0 - e^T P_2 N B_f e - e^T P_2 N B_d d_1) - 2e^T P_2 N B_d d_1 \\
&\leq e^T(-sym(P_2 N B_f) + \beta_1^2 P_2 N (P_2 N)^T + \beta_2^2 P_2 P_2^T)e \\
&\quad + \frac{1}{\beta_1^2} d_1^T B_d^T B_d d_1 + \frac{1}{\beta_2^2} \dot{d}_0^T \dot{d}_0
\end{aligned}
$$

Then, it can be verified that

$$
\begin{aligned}
\dot{V}(x,e,t) &= \dot{V}_1(e,t) + \dot{V}_2(x,t) \\
&\leq x^T(sym(P_1 A + P_1 B_u K) + \alpha_1^2 P_1 P_1^T)x \\
&\quad + e^T(-sym(P_2 N B_f) + \beta_1^2 P_2 N (P_2 N)^T + \beta_2^2 P_2 P_2^T)e \\
&\quad + \frac{1}{\alpha_1^2} \|B_d\|^2 W_1^2 + 2x^T P_1 B_f e \\
&\quad + \frac{1}{\beta_1^2} \|B_d\|^2 W_1^2 + \frac{1}{\beta_2^2} W_0^2 \\
&= \begin{bmatrix} x^T & e^T \end{bmatrix} \Gamma \begin{bmatrix} x \\ e \end{bmatrix} + \frac{1}{\beta_2^2} W_0^2 + \left(\frac{1}{\alpha_1^2} + \frac{1}{\beta_1^2}\right) \|B_d\|^2 W_1^2 \\
&= \begin{bmatrix} x^T & e^T \end{bmatrix} \begin{bmatrix} P_1 & 0 \\ 0 & I \end{bmatrix} \Upsilon \begin{bmatrix} P_1 & 0 \\ 0 & I \end{bmatrix} \begin{bmatrix} x \\ e \end{bmatrix} + C_b
\end{aligned}
$$

where

$$
\Gamma = \begin{bmatrix} \Gamma_{11} & P_1 B_f \\ B_f^T P_1 & \Gamma_{22} \end{bmatrix},
$$

$$
\Gamma_{11} = sym(P_1 A + P_1 B_u K) + \alpha_1^2 P_1 P_1^T,
$$

$$
\Gamma_{22} = -sym(P_2 N B_f) + \beta_1^2 P_2 N (P_2 N)^T + \beta_2^2 P_2 P_2^T,
$$

$$
C_b = \frac{1}{\beta_2^2} W_0^2 + \left(\frac{1}{\alpha_1^2} + \frac{1}{\beta_1^2}\right) \|B_d\|^2 W_1^2, \tag{10.9}
$$

$$
\Upsilon = \begin{bmatrix} P_1^{-1} & 0 \\ 0 & I \end{bmatrix} \Gamma \begin{bmatrix} P_1^{-1} & 0 \\ 0 & I \end{bmatrix} = \begin{bmatrix} \Upsilon_{11} & B_f \\ B_f^T & \Upsilon_{22} \end{bmatrix},
$$

$$\Upsilon_{11} = sym(AP_1^{-1} + B_u KP_1^{-1}) + \alpha_1^2 I,$$

$$\Upsilon_{22} = -sym(Q_2 B_f) + \beta_1^2 Q_2 (Q_2)^T + \beta_2^2 P_2 P_2^T.$$

Denote $Q_1 = P_1^{-1}$ and $R_1 = KP_1^{-1}$, then by using the well-known Schur complement formula, it is shown that $\Theta < 0 \Leftrightarrow \Upsilon < 0$. Denote

$$\Xi = \begin{bmatrix} P_1 & 0 \\ 0 & I \end{bmatrix} \Upsilon \begin{bmatrix} P_1 & 0 \\ 0 & I \end{bmatrix},$$

then $\Theta < 0 \Leftrightarrow \Xi < 0$ holds, which leads to the following conclusion that there exists a positive scalar $\sigma_1 > 0$ such that $\Xi < -\sigma_1 I$. So we can get

$$\dot{V}(x,e,t) < -\sigma_1 \left\| \begin{bmatrix} x \\ e \end{bmatrix} \right\|^2 + C_b,$$

where C_b is defined by (10.9). The conclusion can be obtained. $\qquad\square$

In Theorem 10.1, $\alpha_1 > 0$, $\beta_1 > 0$ and $\beta_2 > 0$ are necessary parameters to get the controller gain and observer gain. We should select appropriate values to get desired attitude control performances.

10.4   SIMULATIONS

In this section, the effectiveness of the present algorithm is demonstrated by numerical simulations, the effect of the elastic vibration is observed and compensated for by DOBC, then fine attitude control is obtained. The composite controller has been applied to the attitude control of a spacecraft with solar paddles. Four bending modes have been considered for the practical spacecraft model at $\omega_1 = 3.17$, $\omega_2 = 7.38$, $\omega_3 = 16.954$, and $\omega_4 = 57.938$ rad/s with damping $\xi_1 = 0.0001$, $\xi_2 = 0.00015$, $\xi_3 = 0.000173$, and $\xi_4 = 0.0001576$, respectively. Because low-frequency modes are generally dominant in a flexible system, in this work, only the first two bending modes are used to represent the displacement of the flexible appendage in the simulation of the system. As an example, in this chapter we try to control the attitude in the pitch channel, where $J = 35.72$ $kg * m^2$ (J is the nominal principal moment of inertia of pitch axis). In addition, $\pm 20\%$ perturbation of the nominal moment of inertial is also considered.

The flexible spacecraft is designed to move in a circular orbit with an altitude of 500 km, then the orbit rate $n = 0.0011$ rad/s. The space environmental disturbance torques acted on the satellite are supposed as follows

$$\begin{cases} T_{dx} &= 4.5 \times 10^{-5}(3\cos nt + 1) \\ T_{dy} &= 4.5 \times 10^{-5}(3\cos nt + 1.5\sin nt) \\ T_{dz} &= 4.5 \times 10^{-5}(3\sin nt + 1) \end{cases}$$

The initial pitch attitude of the spacecraft is

$$\theta = 0.08rad, \dot{\theta} = 0.001rad/s$$

The intermediate design parameters in Theorem 10.1 are $\alpha_1 = 1600$, $\beta_1 = 1$ and $\beta_2 = 1000$, then we can get the parameters of the PD controller

$$K_P = 4.3515, K_D = 17.2488$$

and the observer gain

$$N = \begin{bmatrix} 0 & 1481.3 \end{bmatrix}$$

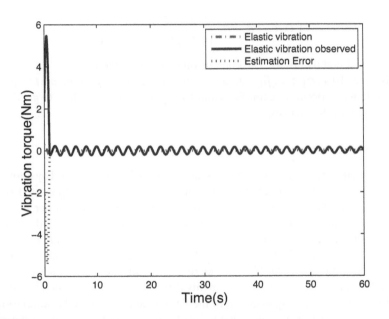

FIGURE 10.3   Time responses of vibration and vibration observed.

In this chapter, $F(2\xi\omega\dot{\eta} + \omega^2\eta)$ is considered as the vibration torque caused by the flexible appendages. Figure 10.3 shows the time responses of elastic vibration, vibration observed and estimation error. Figure 10.4 is obtained by partial amplification of Figure 10.3, and it shows that the vibration from the flexible appendages can be effectively estimated by disturbance observer, where the estimation error is less than 5% of practical elastic vibration. Thus, the effect of the elastic vibration on the rigid hub is reduced to its lowest by feedforward compensation. Figure 10.5 shows that the attitude angle of the spacecraft has fine dynamic response performance under the composite controller. Figure 10.6 is obtained by partial amplification of Figure 10.5, and it shows that the attitude control accuracy is improved at least one order of

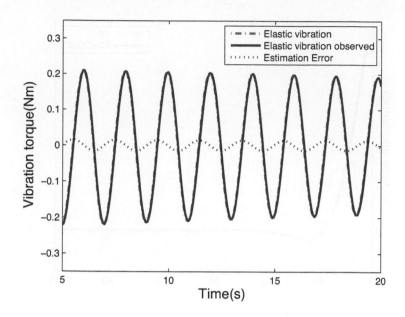

FIGURE 10.4   Vibration estimation error in disturbance observer.

magnitude by composite controller compared with pure PD controller. Figure 10.7 shows that attitude stabilization is significantly improved under composite controller compared with pure PD controller in the presence of flexible vibration.

10.5   CONCLUSION

In this chapter, a new composite control scheme for flexible spacecraft attitude control is discussed in the presence of model uncertainty, elastic vibration and external disturbances. A hybrid controller combining DOBC and PD control is designed for the ACS, in which DOBC can reject the effect of the elastic vibration from the flexible appendages, and a PD controller can effectively perform attitude control for the rigid hub in the presence of multiple disturbances. The closed-loop system stability is proved by stringent analysis when multiple disturbances are taken into consideration.

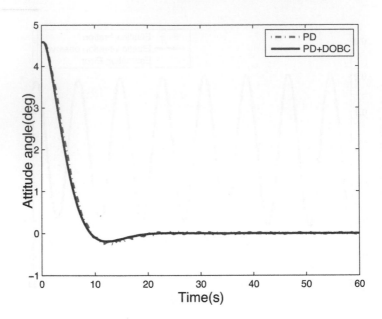

FIGURE 10.5   Time responses of attitude angle.

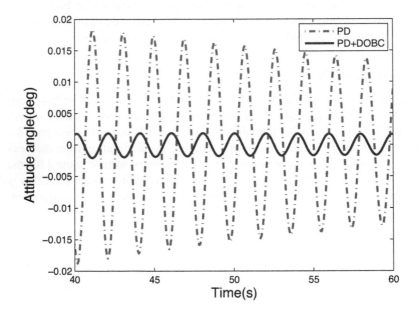

FIGURE 10.6   Pointing precision under different controller.

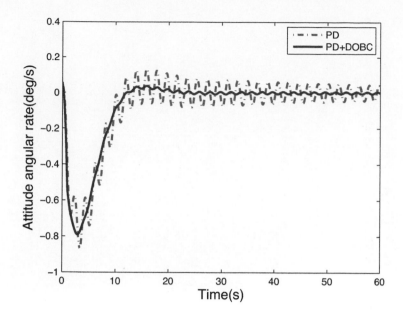

FIGURE 10.7   Time responses of attitude angular velocity.

11 Robust Multi-Objective Initial Alignment for INS with Multiple Disturbances

11.1 INTRODUCTION

Inertial navigation system (INS) has particular advantages and is widely used for the positioning and navigation of vehicles, airplanes, ships, etc. [77]. INS can be classified into gimbaled INS (GINS) and strapdown INS (SINS). One of the key techniques for INS is initial alignment and calibration, which affects navigation accuracy directly. INS mainly includes stationary base alignment and in-flight alignment. Self-alignment schemes have been widely adopted in initial alignment for INS on a stationary base [118, 199]. The purpose of initial alignment of SINS is to determine the coordinate transformation matrix from body frame to computation frame. Then, misalignment angles can be compensated for based on their estimated values. The principle of initial alignment of SINS is similar to that of GINS, where filtering techniques are the key technologies and have been widely used in practice [6].

The problem of state filtering and estimation has been a hot research topic, where the Kalman filter (KF) has been widely adopted as one of the optimal estimators to minimize variance. The standard KF assumes that the system model is exactly known and that the noises are Gaussian noise processes with known statistical properties. Although these cases are generally not satisfied in practical applications, most of the presented results for initial alignment were based on KF and obtained "accepted"performance in practice (see [60, 199] and references therein). However, it has been shown that the performance of the standard KF could be degraded for a related system with model uncertainties and non-Gaussian noises. This has stimulated interest in researching robust and stochastic filtering methods for uncertain and non-Gaussian systems.

In the past decades, following the development of robust control theory, some effective schemes have been devoted to filtering and estimation problems (see [95, 170, 216, 227] and references therein). The advantage of the robust filter in comparison with the Kalman filter is that no statistical assumptions on the noise are required, and the filter is more robust when there exists model uncertainty in an INS. Thus, it has been shown that robust filtering is useful in many practical processes. Among them, the H_2 filtering is to ensure a bound on the H_2 norm from the Gaussian noise signal to the filtering error. The H_∞ performance index is to minimize the H_∞ norm from the norm-bounded noise to the estimation error. In order to guarantee that the filtering system not only demonstrates convergence of estimation error but also has satisfactory dynamic performance, mixed H_2/H_∞ performances have been introduced in filtering problems (see [95, 198] and references therein). This technology has been used to deal with the INS in [198]. Other filtering algorithms having potential applications in initial alignment include intelligent filtering [214] and non-Gaussian filtering [91].

The idea of disturbance attenuation has been applied in most of the existing results for filter design problems, but the known characteristics of some disturbances have not been exactly utilized. In filtering problems, not only the external noise, measurement noise, structure vibration, etc., but also unmodeled dynamics, nonlinear and uncertain dynamics are usually all merged into a disturbance variable. The accuracy of KF and minimum variance filtering cannot satisfy these requirements, since the "merged" disturbance is not Gaussian in most applications. On the other hand, robust filtering methods have a large conservativeness for disturbances described by a norm-bounded variable. In [19], disturbance observer based control (DOBC) is firstly used to design the filter and has shown enhanced anti-disturbance performance.

Generally speaking, to improve the filtering accuracy and enhance the anti-disturbance capability, we need to sufficiently analyze the characteristics of various disturbances based on measurement information. Furthermore, we can formulate the models and describe the different types of multiple disturbances, respectively. As such, new filters with less conservativeness can be provided to estimate and compensate for one part of disturbances. In order to improve the accuracy, both disturbance rejection and attenuation performances should be achieved simultaneously for systems with multiple disturbances.

In this chapter, we analyze the different types of disturbances in an INS and divide them into three parts. One part is modeled to describe inertial sensor drifts, another part is for Gaussian noise and the last is for norm-bounded disturbances, which describe modeling uncertainties, perturbations and non-Gaussian stochastic noises. A mixed multi-objective filter is designed with simultaneous disturbance rejection and attenuation performance for INS with multiple disturbances. The structure of the proposed filter consists of a drift estimator and a mixed H_2/H_∞ filter, where the drift estimator is applied to estimate and reject the inertial sensor drift, and the H_2/H_∞ filter is used to attenuate the Gaussian noises, modeling errors and other bounded external disturbances.

11.2   SINS ERROR MODEL FOR STATIONARY BASE

The process of initial alignment for INS consists of coarse and fine alignment. A coordinate transformation matrix from body frame to computation frame is approximately estimated in the coarse alignment stage. In the fine alignment stage, the small misalignment angles between the reference frame and the true frame are computed accurately and the precise initial transformation matrix is formulated. In this chapter, the local level ENU (East-North-Up) frame is selected as the navigation frame. For the SINS on ground stationary base, positional and velocity errors are ignored. Then, the modified Bar-Itzhack and Berman's SINS error model in the fine alignment stage can be formulated as follows (see [6]).

$$\begin{bmatrix} \dot{\phi}_E(t) \\ \dot{\phi}_N(t) \\ \dot{\phi}_U(t) \end{bmatrix} = F \begin{bmatrix} \phi_E(t) \\ \phi_N(t) \\ \phi_U(t) \end{bmatrix} + C_b^n \begin{bmatrix} \varepsilon_x(t) \\ \varepsilon_y(t) \\ \varepsilon_z(t) \end{bmatrix} + B_1 \omega_1(t) + B_2 \omega_2(t) \qquad (11.1)$$

where ϕ and ε are the attitude errors and gyroscope drifts, respectively. The subscript E, N and U represent east, north, up of the navigation frame n, and the subscript x, y and z denote right, front, up of the body frame b, respectively. The matrix F is defined as

$$F = \begin{bmatrix} 0 & \Omega_U & -\Omega_N \\ -\Omega_U & 0 & 0 \\ \Omega_N & 0 & 0 \end{bmatrix} \qquad (11.2)$$

where Ω_U and Ω_N are the up and north components of earth rate, respectively. C_b^n is the coordinate transformation matrix relating the body frame b with navigation frame n, which can be computed by the roll, pitch and yaw angles of the vehicle.

$$C_b^n = \{C_{ij}\}_{i,j=1,2,3} \qquad (11.3)$$

$\omega_1(t)$ is Gaussian stochastic noise of SINS. $\omega_2(t)$ represents a norm-bounded disturbance, which results from the non-Gaussian noise, modeling error and uncertainties in the SINS. B_1 and B_2 represent the known coefficient matrices of the system with suitable dimensions.

In frame n, the local gravity acceleration g and earth rate w_{ie} can be described by

$$g^n = [\, 0 \quad 0 \quad -g \,]^T \qquad (11.4)$$

and

$$w_{ie}^n = [\, 0 \quad \Omega_N \quad \Omega_U \,]^T \qquad (11.5)$$

The superscript n denotes the navigation frame. The gyroscope is applied to measure the earth rate w_{ie} and an accelerometer to measure the local gravity acceleration g for SINS on a ground stationary base. Therefore, the level accelerometer outputs $f_E(t)$, $f_N(t)$ and east gyro output $w_E(t)$ are chosen as measured signals during the stationary alignment [61]

$$\begin{bmatrix} f_E(t) \\ f_N(t) \\ w_E(t) \end{bmatrix} = H \begin{bmatrix} \phi_E(t) \\ \phi_N(t) \\ \phi_U(t) \end{bmatrix} + D^b \omega(t) + D_1 \omega_1(t) + D_2 \omega_2(t) \qquad (11.6)$$

where the output matrix is denoted by

$$H = \begin{bmatrix} 0 & g & 0 \\ -g & 0 & 0 \\ 0 & -\Omega_U & \Omega_N \end{bmatrix} \qquad (11.7)$$

and

$$D^b = \begin{bmatrix} 0 & 0 & 0 & C_{11} & C_{12} \\ 0 & 0 & 0 & C_{21} & C_{22} \\ C_{11} & C_{12} & C_{13} & 0 & 0 \end{bmatrix} \qquad (11.8)$$

C_{ij} $(i, j = 1, 2, 3)$ is defined in (11.4), D_1 and D_2 represent the known coefficient matrices with suitable dimensions. The inertial sensor bias variable is denoted by

$$\omega^T(t) = [\ \varepsilon_x(t) \quad \varepsilon_y(t) \quad \varepsilon_z(t) \quad \nabla_x(t) \quad \nabla_y(t)\ ]$$

where ∇ represents the accelerometer biases. In this chapter, the inertial sensor bias variable is supposed to satisfy a first-order Gaussian Markov process (see [177]), which can be formulated as

$$\dot{\omega}(t) = W\omega(t) + B_3 \omega_1(t) \qquad (11.9)$$

where

$$W = diag\{-\frac{1}{\tau_1}, \cdots, -\frac{1}{\tau_5}\}$$

and τ_i $(i = 1, \cdots, 5)$ are correlation times of the Markov process. B_3 is also a known matrix.

Remark 11.1 The specific force and east gyro outputs were selected as the measurement signal in [61]. In [60, 177, 199], the specific force errors were once integrated to obtain the velocity errors, which were adopted as the measurement variable. In this chapter, we choose the specific force and east gyro output as the measurement signals. Different from previous work in [60, 198, 199] assuming the drifts to be random constants, we use (11.9) to represent the inertial sensor drifts. Gaussian Markov processes are useful in many engineering applications since they can describe many physical random processes with good approximation [177].

Remark 11.2 In most of the previous works, the disturbance of SINS was supposed to be a Gaussian noise signal (see [60, 199]), which was not satisfied in many practical processes. For the SINS error model (11.1), Gaussian noise $\omega_1(t)$ is caused by inertial sensors, and the inertial sensor drift $\omega(t)$ is supposed to be a Markov process. Machine vibration, modeling error and environment disturbance of SINS can be equal to a norm-bounded disturbance $\omega_2(t)$. Consequently, the concerned SINS is a class of linear systems with multiple disturbances.

The relationships of gyro and accelerometer outputs between the navigation frame and the body frame can be denoted as follows

$$C_b^n [\ \varepsilon_x \quad \varepsilon_y \quad \varepsilon_z\ ]^T = [\ \varepsilon_E \quad \varepsilon_N \quad \varepsilon_U\ ]^T \qquad (11.10)$$

$$C_b^n [\ \nabla_x \quad \nabla_y \quad \nabla_z\ ]^T = [\ \nabla_E \quad \nabla_N \quad \nabla_U\ ]^T \qquad (11.11)$$

Based on (11.10), (11.11), and Lyapunov transformation in [41], the time-varying SINS error Equations (11.1) and (11.6) can be further formulated as follows

$$
\begin{cases}
\begin{bmatrix} \dot{\phi}_E(t) \\ \dot{\phi}_N(t) \\ \dot{\phi}_U(t) \end{bmatrix} = F \begin{bmatrix} \phi_E(t) \\ \phi_N(t) \\ \phi_U(t) \end{bmatrix} + \begin{bmatrix} \varepsilon_E(t) \\ \varepsilon_N(t) \\ \varepsilon_U(t) \end{bmatrix} \\
\qquad\qquad + B_1 \omega_1(t) + B_2 \omega_2(t) \\
\begin{bmatrix} f_E(t) \\ f_N(t) \\ w_E(t) \end{bmatrix} = H \begin{bmatrix} \phi_E(t) \\ \phi_N(t) \\ \phi_U(t) \end{bmatrix} + \begin{bmatrix} \nabla_E(t) \\ \nabla_N(t) \\ \varepsilon_E(t) \end{bmatrix} \\
\qquad\qquad + D_1 \omega_1(t) + D_2 \omega_2(t)
\end{cases}
\tag{11.12}
$$

The transformed system model (11.12) of SINS is the same as the model of GINS in the form. The discretization form of system (11.12) can be concluded as follows

$$
\begin{cases}
x(k+1) = Ax(k) + B\omega^n(k) + u(k) + B_1\omega_1(k) + B_2\omega_2(k) \\
y(k) = Cx(k) + D\omega^n(k) + u(k) + D_1\omega_1(k) + D_2\omega_2(k)
\end{cases}
\tag{11.13}
$$

where the system state is

$$
x^T(k) = [\ \phi_E(k) \quad \phi_N(k) \quad \phi_U(k)\ ],
$$

system output variable is

$$
y^T(k) = [\ f_E(k) \quad f_N(k) \quad w_E(k)\ ],
$$

$u(k)$ is the control input for calibration and compensation. And

$$
\omega^n(k) = [\ \varepsilon_E(k) \quad \varepsilon_N(k) \quad \varepsilon_U(k) \quad \nabla_E(k) \quad \nabla_N(k)\ ]^T
$$

is the inertial sensor drift in the navigation frame n.

After the SINS error model is established, the initial alignment for SINS on a ground stationary base can be transformed into an anti-disturbance filter design problem for linear time invariant systems with multiple disturbances.

11.3   ROBUST MULTI-OBJECTIVE FILTER DESIGN

In this section, the intention is to design a robust multi-objective filter in the presence of multiple disturbances for related linear discrete systems. Drift estimations will be applied to reject the inertial sensor drifts. H_∞ optimization technique will be adopted to attenuate the norm-bounded disturbances and H_2 performance is to optimize the error variance. Accordingly, the proposed multi-objective filter has both disturbance rejection and attenuation performance. For this purpose, we design an inertial sensor drift estimator as follows

$$
\widehat{\omega^n}(k+1) = W\widehat{\omega^n}(k) + K[y(k) - \hat{y}(k)]
\tag{11.14}
$$

where $\widehat{\omega^n}(k)$ is the estimation of inertial sensor drifts $\omega^n(k)$. Based on (11.14), the mixed H_2/H_∞ filter can be constructed as

$$\begin{cases} \widehat{x}(k+1) & = & A\widehat{x}(k) + u(k) + u_{c1}(k) + L[y(k) - \widehat{y}(k)] \\ \widehat{y}(k) & = & C\widehat{x}(k) + u(k) + u_{c2}(k) \end{cases} \tag{11.15}$$

where $\widehat{x}(k)$ is the estimation of state $x(k)$, $\widehat{y}(k)$ is the estimation of output signal $y(k)$. The control inputs $u_{c1}(k) = B\widehat{\omega^n}(k)$ and $u_{c2}(k) = D\widehat{\omega^n}(k)$ are applied to compensate and calibrate the inertial sensor drifts. K and L are the filter gains to be determined.

Denoting

$$\widetilde{x}(k) = x(k) - \widehat{x}(k),$$

and

$$\widetilde{\omega^n}(k) = \omega^n(k) - \widehat{\omega^n}(k),$$

the estimation error system yields

$$\begin{cases} \widetilde{x}(k+1) & = & (A - LC)\widetilde{x}(k) + (B - LD)\widetilde{\omega^n}(k) \\ & & + (B_1 - LD_1)\omega_1(k) + (B_2 - LD_2)\omega_2(k) \\ \widetilde{\omega^n}(k+1) & = & (W - KD)\widetilde{\omega^n}(k) + (B_3 - KD_1)\omega_1(k) \\ & & - KC\widetilde{x}(k) - KD_2\omega_2(k) \end{cases} \tag{11.16}$$

Definition 11.1 The H_∞ reference output is defined as

$$z_\infty(k) = C_{\infty 1}\widetilde{x}(k) + C_{\infty 2}\widetilde{\omega^n}(k) \tag{11.17}$$

and the H_2 reference output is defined as

$$z_2(k) = C_{21}\widetilde{x}(k) + C_{22}\widetilde{\omega^n}(k) \tag{11.18}$$

where $C_{\infty 1}$, $C_{\infty 2}$, C_{21} and C_{22} are selected weighting matrices.

Combining Equations (8.17) and (8.18) with estimation error Equation (8.16) yields

$$\begin{cases} \begin{bmatrix} \widetilde{x}(k+1) \\ \widetilde{\omega^n}(k+1) \end{bmatrix} & = & \overline{A} \begin{bmatrix} \widetilde{x}(k) \\ \widetilde{\omega^n}(k) \end{bmatrix} + \overline{B_1}\omega_1(k) + \overline{B_2}\omega_2(k) \\ z_\infty(k) & = & \overline{C_\infty} \begin{bmatrix} \widetilde{x}(k) \\ \widetilde{\omega^n}(k) \end{bmatrix} \\ z_2(k) & = & \overline{C_2} \begin{bmatrix} \widetilde{x}(k) \\ \widetilde{\omega^n}(k) \end{bmatrix} \end{cases} \tag{11.19}$$

where

$$\overline{A} = \begin{bmatrix} A - LC & B - LD \\ -KC & W - KD \end{bmatrix},$$

$$\overline{B_1} = \begin{bmatrix} B_1 - LD_1 \\ B_3 - KD_1 \end{bmatrix}, \overline{B_2} = \begin{bmatrix} B_2 - LD_2 \\ -KD_2 \end{bmatrix},$$

and

$$\overline{C_2} = [\ C_{21} \quad C_{22}\ ], \overline{C_\infty} = [\ C_{\infty 1} \quad C_{\infty 2}\ ].$$

At this stage, the objective is to find K and L such that system (11.19) is stable and satisfied with the mixed performance index. The following result provides a new robust initial alignment method based on LMIs (linear matrix inequalities).

Before proceeding further, we consider a general discrete system

$$\begin{cases} \xi(k+1) &= A\xi(k) + B_1 d_1(k) + B_2 d_2(k) \\ J_\infty(k) &= L_\infty \xi(k) + D_2 d_2(k) \\ J_2(k) &= L_2 \xi(k) + D_1 d_1(k) \end{cases} \tag{11.20}$$

where $\xi(k)$ is the state variable, $d_1(k)$ is the Gaussian noise, and $d_2(k)$ is a norm-bounded disturbance. H_∞ reference output is supposed to be $J_\infty(k)$ and H_2 reference output is assumed to be $J_2(k)$.

The following lemma will be used in the proof of our main results in this section (see [95, 216] for details).

Lemma 11.1

If for weighting parameters $r_1 > 0, r_2 > 0$ and weighing matrices L_2, L_∞, there exist matrices $Z > 0, P > 0$ and parameter $\gamma_\infty > 0$ satisfying

$$\min r_1 Tr(Z) + r_2 \gamma_\infty$$

$$\begin{bmatrix} -Z & B_1^T P & D_1^T \\ * & -P & 0 \\ * & * & -I \end{bmatrix} < 0$$

$$\begin{bmatrix} -P & A^T P & L_2^T \\ * & -P & 0 \\ * & * & -I \end{bmatrix} < 0$$

$$\begin{bmatrix} -P & 0 & A^T P & L_\infty^T \\ * & -\gamma_\infty I & B_2^T P & D_2^T \\ * & * & -P & 0 \\ * & * & * & -I \end{bmatrix} < 0$$

then the system (11.20) is stable, H_2 norm of system from $d_1(k)$ to $J_2(k)$ is less than $Tr(Z)$ and H_∞ norm of system from $d_2(k)$ to $J_\infty(k)$ is less than γ_∞.   ∎

Based on the above Lemma, we have the following filter design criterion for initial alignment of INS.

Theorem 11.1

If for weighting parameters $r_1 > 0, r_2 > 0$ and weighting matrices $C_{\infty 1}, C_{\infty 2}, C_{21}, C_{22}$, there exist matrices $G_1, G_2, \Pi > 0, P_i > 0 \ (i = 1, 2, 3, 4)$ and R_1, R_2 satisfying

$$\min r_1 Tr(\Pi) + r_2 \gamma \tag{11.21}$$

$$\begin{bmatrix} \Xi_3 & 0 & G_1 B_1 - R_1 D_1 \\ * & \Xi_4 & G_2 B_3 - R_2 D_1 \\ * & * & -\Pi \end{bmatrix} < 0 \tag{11.22}$$

$$\begin{bmatrix} \Xi_3 & 0 & \Xi_{13} & \Xi_{14} & 0 \\ * & \Xi_4 & -R_2 C & \Xi_{24} & 0 \\ * & * & -P_3 & 0 & C_{21}^T \\ * & * & * & -P_4 & C_{22}^T \\ * & * & * & * & -I \end{bmatrix} < 0 \tag{11.23}$$

$$\begin{bmatrix} \Xi_1 & 0 & \Xi_{13} & \Xi_{14} & \Xi_{15} & 0 \\ * & \Xi_2 & -R_2 C & \Xi_{24} & \Xi_{25} & 0 \\ * & * & -P_1 & 0 & 0 & C_{\infty 1}^T \\ * & * & * & -P_2 & 0 & C_{\infty 2}^T \\ * & * & * & * & -\gamma I & 0 \\ * & * & * & * & * & -I \end{bmatrix} < 0 \tag{11.24}$$

where

$$\Xi_1 = P_1 - G_1 - G_1^T, \quad \Xi_2 = P_2 - G_2 - G_2^T,$$

$$\Xi_3 = P_3 - G_1 - G_1^T, \quad \Xi_4 = P_4 - G_2 - G_2^T,$$

$$\Xi_{13} = G_1 A - R_1 C, \quad \Xi_{14} = G_1 B - R_1 D,$$

$$\Xi_{15} = G_1 B_2 - R_1 D_2, \quad \Xi_{24} = G_1 W - R_2 D, \quad \Xi_{25} = -R_2 D_2.$$

then with gains $L = G_1^{-1} R_1$ and $K = G_2^{-1} R_2$ error system (11.19) is stable, H_2 norm of system from $\omega_1(k)$ to $z_2(k)$ is less than $Tr(\Pi)$ and H_∞ norm of system from $\omega_2(k)$ to $z_\infty(k)$ is less than γ. ∎

Proof. Applying Lemma 11.1 to system (11.19), it can be shown that

$$\min r_1 Tr(\Pi) + r_2 \gamma$$

$$\begin{bmatrix} -\Pi & B_1^T P \\ * & -P \end{bmatrix} < 0$$

$$\begin{bmatrix} -P & \overline{A}^T P & \overline{C_2}^T \\ * & -P & 0 \\ * & * & -I \end{bmatrix} < 0$$

$$\begin{bmatrix} -P & 0 & \overline{A}^T P & \overline{C_\infty}^T \\ * & -\gamma I & \overline{B_2}^T P & 0 \\ * & * & -P & 0 \\ * & * & * & -I \end{bmatrix} < 0$$

Using the optimization synthesis method in [165], the Lyapunov and the system matrices can be separated, which will reduce the complexity of the proposed algorithm. The above-mentioned LMIs can be further concluded as follows

$$\min r_1 Tr(\Pi) + r_2 \gamma$$

$$\begin{bmatrix} -\Pi & G\overline{B_1} \\ * & \overline{P_2} - G - G^T \end{bmatrix} < 0$$

$$\begin{bmatrix} \overline{P_2} - G - G^T & G\overline{A} & 0 \\ * & -\overline{P_2} & \overline{C_2}^T \\ * & * & -I \end{bmatrix} < 0$$

$$\begin{bmatrix} \overline{P_1} - G - G^T & G\overline{A} & G\overline{B_2} & 0 \\ * & -\overline{P_1} & 0 & \overline{C_\infty}^T \\ * & * & -\gamma I & 0 \\ * & * & * & -I \end{bmatrix} < 0$$

From the definition of weighting matrices of (11.19), we denote

$$G = \begin{bmatrix} G_1 & 0 \\ 0 & G_2 \end{bmatrix}, \overline{P_1} = \begin{bmatrix} P_1 & 0 \\ 0 & P_2 \end{bmatrix},$$

$$\overline{P_2} = \begin{bmatrix} P_3 & 0 \\ 0 & P_4 \end{bmatrix}, G_1 L = R_1, G_2 K = R_2.$$

Then, LMIs (11.21-11.24) can be verified. This completes the proof.   □

Theorem 11.1 provided a robust initial alignment design method based on the mixed multi-objective optimization principle.

11.4   SIMULATION EXAMPLES

In this section, the stationary base alignments of SINS in laboratory and airborne environment are considered, respectively. Simulation conditions:

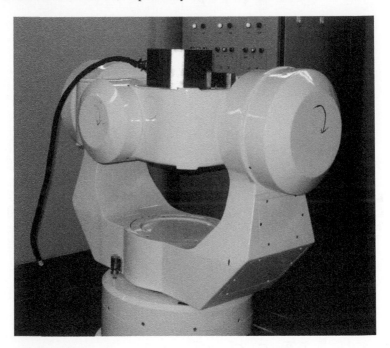

FIGURE 11.1   INS in the turn table.

1. Local latitude is $39.9degN$;
2. Initial misalignment angles ϕ_U, ϕ_E, ϕ_N are chosen as $1deg$, $0.1deg$ and $0.1deg$, respectively;
3. Gyroscopes: the relative times of Markov process are $\tau_1 = \tau_2 = \tau_3 = 3600s$ and random drift is $0.05deg/hr$;
4. Accelerometers: the relative times of Markov process are $\tau_4 = \tau_5 = 1800s$ and random drift is $50\mu g$.

The correlative times for the random bias are decided based on the properties of the selected inertial sensors by testing. The coefficient matrices of error model (11.13)

TABLE 11.1

The STD of estimation error of misalignment angles ($10^{-6}rad$) in the laboratory.

Method	leveling x	leveling y	azimuth
In this chapter	0.522605	0.300066	7.017749
KF	1.353420	8.195616	371.979119

are given by

$$A = 10^{-4} \times \begin{bmatrix} 0 & 0.467752 & -0.559426 \\ -0.467752 & 0 & 0 \\ 0.559426 & 0 & 0 \end{bmatrix},$$

$$C = \begin{bmatrix} 0 & 9.801609 & 0 \\ -9.801609 & 0 & 0 \\ 0 & -0.000047 & 0.000056 \end{bmatrix},$$

$$B = \begin{bmatrix} I_3 & 0_{3\times2} \end{bmatrix}, B_1 = 0,$$

$$D = \begin{bmatrix} 0 & 0 & 0 & 1 & 0 \\ 0 & 0 & 0 & 0 & 1 \\ 1 & 0 & 0 & 0 & 0 \end{bmatrix}.$$

11.4.1 SINS IN LABORATORY ENVIRONMENT

Firstly, we consider the SINS data obtained in an experiment in the laboratory (see Figure 11.1). Then, the parameter matrices of the different types of disturbances are formulated as follows

$$B_2 = diag\{\frac{0.00001\pi}{18000}, \frac{0.00001\pi}{18000}, \frac{0.0001\pi}{180}\},$$

$$B_3 = diag\{\frac{0.0001\pi}{648000}, \frac{0.0001\pi}{648000}, \frac{0.0001\pi}{648000}, 10^{-7}g, 10^{-7}g\},$$

$$D_1 = \begin{bmatrix} 10^{-3}g & 0 & 0 \\ 0 & 10^{-3}g & 0 \\ 0 & 0 & \frac{1\pi}{648000} \end{bmatrix},$$

$$D_2 = 0.001D_1.$$

The parameter matrices of KF are selected as follows. The initial mean square error matrix is

$$P(0) = diag\{(0.1deg)^2, (0.1deg)^2, (1deg)^2, (0.1deg/hr)^2,$$

$$(0.1deg/hr)^2, (0.1deg/hr)^2, (100\mu g)^2, (100\mu g)^2\}$$

The variance matrix of system noise is

$$Q = diag\{(0.05deg/hr)^2, (0.05deg/hr)^2, (0.05deg/hr)^2\},$$

and variance matrix of measurement noise is

$$R = diag\{(50\mu g)^2, (50\mu g)^2, (0.05deg/hr)^2\}.$$

For the robust multi-objective filter, the H_∞ reference output weighting matrices are selected as

$$C_{\infty 1} = \begin{bmatrix} 0.1 & 0.1 & 0.01 \end{bmatrix},$$

$$C_{\infty 2} = \begin{bmatrix} 0 & 0 & 0 & 0 & 0 \end{bmatrix}$$

The H_2 reference output weighting matrices are selected as

$$C_{21} = \begin{bmatrix} 10^{-6}g & 10^{-6}g & 10^{-5}g \end{bmatrix},$$

$$C_{22} = \begin{bmatrix} \frac{0.01\pi}{180} & \frac{0.01\pi}{180} & \frac{0.01\pi}{180} & \frac{0.1\pi}{180} & \frac{0.1\pi}{180} \end{bmatrix}$$

In the laboratory, the SINS is mainly affected by inertial sensors drift, Gaussian measurement noise, and linearization error. To heighten the H_2 performance in the mixed multi-objective filter, we select the H_2 weighting parameter $r_1 = 100$, and H_∞ weighting parameter $r_2 = 1$. It can be solved via LMIs related to (11.21-11.24) that disturbance attenuation value $\gamma = 2.146201 \times 10^{-9}$ and the gain of robust multi-objective filter is

$$L = \begin{bmatrix} -0.000048 & -0.000093 & -0.101043 \\ 0.000076 & -0.000021 & 0.012468 \\ 0.001394 & 0.000454 & 2.002991 \end{bmatrix},$$

the gain of drift estimator is

$$K = 10^{-2} \times \begin{bmatrix} -0.000001 & -0.000000 & -0.000136 \\ 0.000002 & -0.000001 & 0.000419 \\ 0.000000 & 0.000001 & 0.000132 \\ 0.000189 & 0.000385 & 1.023734 \\ -0.000408 & -0.000910 & -0.562703 \end{bmatrix}.$$

The estimation errors of three misalignment angles are described in Figures 11.2-11.4, where the solid lines represent the estimation errors by robust multi-objective filter in this chapter and dash-dot lines denote the estimation errors based on KF. It is shown that the misalignment angle estimation errors of robust multi-objective filter are less than KF in Figures 11.2-11.4. The composition of standard deviations (STDs) of three misalignment angle estimation errors (after the 50th second) in the laboratory between robust multi-objective filter in this chapter and KF is shown in Table 11.1. From Table 11.1, the accuracy of the proposed robust filter is higher than KF for the considered systems with multiple disturbances.

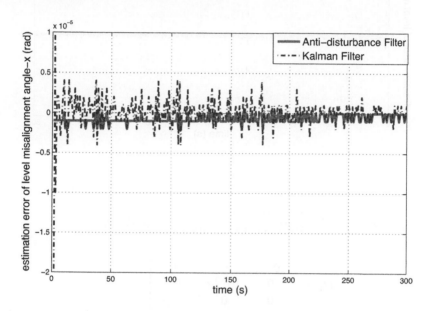

FIGURE 11.2   Estimation errors of ϕ_x in the laboratory.

11.4.2   SINS IN AIRBORNE ENVIRONMENT

Secondly, the stationary base initial alignment of airborne SINS is investigated. For an airborne SINS, there unavoidably exist machine vibrations, modeling errors and outdoor environmental disturbances. These unknown and uncertain variables can be merged into $\omega_2(t)$ as a norm-bounded disturbance for the airborne SINS, where

$$\omega_2^T(t) = \begin{bmatrix} 0.001 & 0.001 & 0.0001 \end{bmatrix} * sign(sin(100t))$$

We insert the non-Gaussian disturbance $\omega_2(t)$ into the SINS data measured in the laboratory. Then, the parameter matrices of the different types of disturbance are formulated as follows.

$$B_2 = diag\{\frac{0.001\pi}{180}, \frac{0.001\pi}{180}, \frac{0.002\pi}{180}\},$$

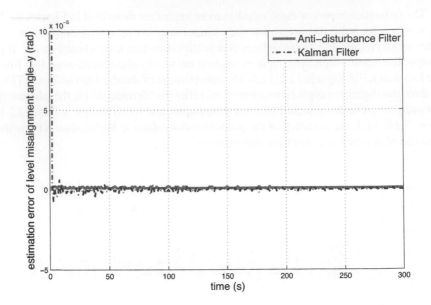

FIGURE 11.3   Estimation errors of ϕ_y in the laboratory.

$$B_3 = diag\{\frac{0.001\pi}{648000}, \frac{0.001\pi}{648000}, \frac{0.001\pi}{648000}, 10^{-6}g, 10^{-6}g\},$$

$$D_1 = \begin{bmatrix} 10^{-3}g & 0 & 0 \\ 0 & 10^{-3}g & 0 \\ 0 & 0 & \frac{1\pi}{648000} \end{bmatrix},$$

$$D_2 = D_1.$$

To heighten the H_∞ performance in the mixed multi-objective filter, we select $r_1 = 1$ and $r_2 = 100$, it can be solved via LMIs related to (11.21-11.24) that $\gamma =$

TABLE 11.2

The STD of estimation error of misalignment angles ($10^{-5}rad$) for airborne SINS.

Method	leveling x	leveling y	azimuth
In this chapter	0.462194	2.164209	24.666823
KF	7.669007	6.339486	730.335094

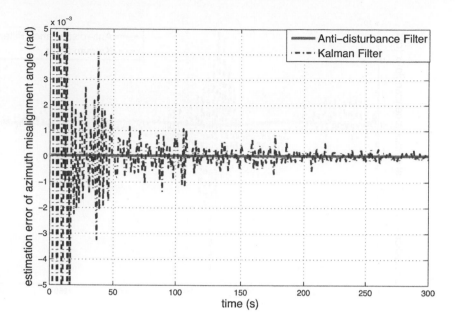

FIGURE 11.4   Estimation errors of ϕ_z in the laboratory.

1.009291×10^{-8} and the gain of robust multi-objective filter is

$$
L = \begin{bmatrix} 0.001439 & -0.001643 & 0.023710 \\ 0.018859 & 0.001828 & -0.516635 \\ 0.002364 & 0.000282 & 6.985335 \end{bmatrix}
$$

the gain of drift estimator is

$$
K = 10^{-3} \times \begin{bmatrix} 0.000001 & 0.000001 & -0.005311 \\ -0.000000 & 0.000001 & -0.000044 \\ 0.000003 & 0.000001 & 0.003083 \\ -0.093282 & -0.022467 & 4.444655 \\ 0.097609 & 0.013189 & -4.596511 \end{bmatrix}.
$$

In Figures 11.5-11.7, the misalignment angle estimation errors based on robust multi-objective filter and KF are demonstrated, where the robust multi-objective filter is more accurate than the counterpart of KF. It can be seen that the proposed robust multi-objective filter has a good ability for initial alignment of SINS with multiple disturbances. The composition of standard deviations of three misalignment angle estimation errors (after the 50th second) for the airborne SINS between robust multi-objective filter and KF is shown in Table 11.2. From Table 11.2, the accuracy of the proposed robust filter is higher than KF for the considered systems with multiple disturbances.

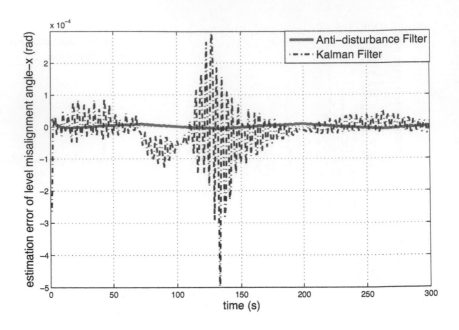

FIGURE 11.5   Estimation errors of ϕ_x for airborne SINS.

11.5   CONCLUSION

In this chapter, the problem of initial alignment for INS is investigated. Different from most previous works which only focus on Gaussian noises (or other single disturbances), an INS error model is constructed with multiple disturbances. The multiple disturbances are supposed to include three parts, which are inertial sensor drifts, Gaussian noise and norm-bounded disturbances. Initial alignment for INS is transformed into a robust filter design problem for systems with multiple disturbances. An anti-disturbance filter is designed for the considered INS with enhanced disturbance rejection and attenuation performance.

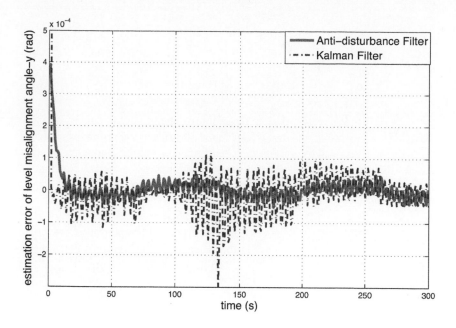

FIGURE 11.6   Estimation errors of ϕ_y for airborne SINS.

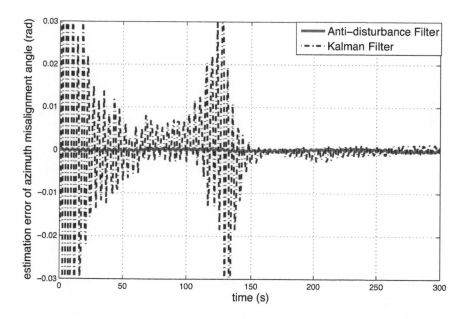

FIGURE 11.7   Estimation errors of ϕ_z for airborne SINS.

12 Nonlinear Initial Alignment for INS with Multiple Disturbances

12.1 INTRODUCTION

An inertial navigation system (INS) is a navigation aid that utilizes motion sensors (accelerometers), rotation sensors (gyroscopes) and a navigation processor to continuously calculate via dead reckoning the position, velocity, and orientation of a moving object. The principle of inertial navigation is simple. With precise knowledge of the initial position and attitude of a vehicle, at any time, single and double integration of the accelerometer outputs will provide velocity and position information, respectively, and the attitude can be obtained by integrating rotational rated from gyro measurements ([156]). In the process of calculating navigation parameters, attitude errors accumulate with time extendibility and induce other navigation errors because the attitude angle terms are integrated for velocity and position. Therefore, one of the most important problems for INS is initial alignment and calibration, which affects the navigation accuracy directly. The purpose of initial alignment of strapdown INS (SINS) is to determine the attitude matrix from body frame to navigation frame and to compensate for misalignment angles based on estimated values. The principle of initial alignment of SINS is similar to that of gimbaled INS (GINS), in both of which filtering is widely utilized as one of the key technologies.

In modern field of control, the problem of state estimation has been one of the hot topics and there have been many works following Kalman filtering (KF) (in a stochastic framework) and Luenberger observers (in a deterministic framework). The KF is one of the optimal linear estimators to minimize error variance, while for nonlinear systems, the extended Kalman filter (EKF) and unscented Kalman filter (UKF) have also been provided. Up to now, Kalman type filtering methods (including KF, EKF and UKF) were effectively used in most of the existing results for the initial alignment problems (see, e.g., [1, 60, 77, 118, 139, 177, 178, 199]). However, the Kalman type filters require the exact plant model and noise with known statistical properties, which are often not satisfied in practical processes. The performance of Kalman type filters could be degraded when the concerned system displays model uncertainties, non-Gaussian noises and other types of disturbances. In order to overcome the drawbacks of KF, robust filtering and estimation problems have been studied in the past decade. Different from Kalman type filters, no statistical assumptions concerning the noises are required and the norm-bounded uncertainties can be dealt with. Among the robust filtering methods, the H_∞ and H_2 filtering are widely applied to state observation, parameter estimation and fault diagnosis problems (see [91, 95, 55], and references

therein). Other new effective filtering methodologies include unknown input estimation (see [64, 71]) and non-Gaussian stochastic filtering (see [91]). Up to now, most
robust filtering has not been effectively applied to solve the initial alignment problem
due to the special characteristics and complexity of INS.

There are two fundamental types of alignment process: stationary base alignment
and in-flight alignment. Self-alignment schemes have been widely adopted for stationary base alignment, where gyroscopes are applied to measure the earth rate and
accelerometers are used to measure the local gravity acceleration ([21, 61, 199]). For
in-flight alignment, the process is to estimate and calibrate platform error angles with
external aiding position and velocity information from GPS or other sensors. The
design of in-flight alignment is mainly based on multi-sensor fusion technologies,
which will be the main trend in navigation system research and development (see
[131, 226, 230]). The low-cost INS based on the micro electro mechanical system
(MEMS) has become a hot research issue with the advantages of small size, light-
weight, low-power consumption, low cost and so on. For low-cost INS, the initial
alignment is still a challenging issue because of the high noise of low-cost inertial
sensors. The initial position and velocity can be easily obtained from GPS or other
aiding sensors ([99, 156]). To transfer information between different sensors that are
located apart from each other, transfer alignment is the process of simultaneously ini-
tializing and calibrating a slave INS using data from master navigation system ([78]).
In addition, two obstacles still exist to influence filtering performance for stationary
base alignment. First, most of the previous alignment methods focused on the small
attitude error case, so linear INS error models can be established and attitude can be
determined coarsely (see [21, 60, 118, 139]). In practice, heading error is much larger
than leveling error, so the linearization of heading error will lead to considerable modeling errors. That is, nonlinear dynamics should be considered for higher precision
of INS. Second, the plant models are supposed to have one single type of disturbance
(either Gaussian variable or norm-bounded variable), for which Kalman type filtering
and robust filtering can be applied. However, along with the development of sensor
technology and data processing, multiple disturbances exist in most practical systems
and one can formulate them into different mathematical descriptions after modeling
analysis and error analysis. For example, inertial sensor drifts can be described by a
Gaussian Markov process as discussed in [177], for which either KF or robust filtering may be inappropriate. In fact, circuit noises, external disturbances and unmodeled
dynamics can be modeled by different formulations including exosystems, harmonic
signals and norm-bounded variables ([19, 21, 85, 201]).

In this chapter, we attempt to present a robust filtering algorithm for the initial
alignment problem of nonlinear INS. A new nonlinear INS error model with multiple disturbances is established, where the nonlinear dynamics and different types of
disturbances in INS are analyzed and modeled, respectively. Concretely, the inertial
sensor drift is described by a Gaussian Markov process. The modeling uncertainties,
perturbations and non-Gaussian stochastic noises are merged into a norm-bounded
variable. With the known characteristics of disturbances, a robust anti-disturbance filter is designed with simultaneous disturbance rejection and attenuation performance

for nonlinear INS. In the proposed scheme, the sensor bias estimations are applied to reject inertial sensor drift, while H_∞ performance is adopted to attenuate norm-bounded uncertain disturbances and the guarantee cost index is applied to optimize estimation error.

12.2 NONLINEAR INS ERROR MODELING

An INS is a dead reckoning navigation system, which comprises a set of inertial sensors and a navigation computer. Theoretically, single and double integration of the gyro and accelerometer outputs will provide velocity and position information. Alignment is the process whereby the orientation of the axes of an INS is determined with respect to a reference coordinate system. The orientation is essential to obtain acceleration in the navigation frame and then to evaluate the single and double integration for velocity and position determination ([156]). The objective of initial alignment for SINS is to determine an attitude matrix from body frame b to navigation frame n (see the 3D diagram plotted in Figure 12.1). We take no account of the linearization of azimuth misalignment angle, because heading error is much larger than leveling error in many practical processes. In this chapter, the local level ENU (East-North-Up) frame is selected as the navigation frame. For an INS on a ground stationary base, the local position of a vehicle can be known in advance. Therefore, the positional and velocity errors are ignored. Then, the modified SINS large azimuth misalignment error model can be formulated as follows (see [178])

$$\dot{\phi} = (I - C_n^t)w_{ie}^n + C_b^n \varepsilon^b \tag{12.1}$$

where $\phi = \begin{bmatrix} \phi_E(t) & \phi_N(t) & \phi_U(t) \end{bmatrix}^T$ and $\varepsilon^b = \begin{bmatrix} \varepsilon_x(t) & \varepsilon_y(t) & \varepsilon_z(t) \end{bmatrix}^T$ are the misalignment angles and gyroscope random bias, respectively. C_n^t is the transformation matrix between the navigation frame n and computation frame t, which can be further described by

$$C_n^t = \begin{bmatrix} cos\phi_U(t) & sin\phi_U(t) & -\phi_N(t) \\ -sin\phi_U(t) & cos\phi_U(t) & \phi_E(t) \\ \overline{\phi_N}(t) & \overline{\phi_E}(t) & 1 \end{bmatrix} \tag{12.2}$$

where

$$\overline{\phi_E}(t) = \phi_N(t)sin\phi_U(t) - \phi_E(t)cos\phi_U(t)$$

and

$$\overline{\phi_N}(t) = \phi_E(t)sin\phi_U(t) + \phi_N(t)cos\phi_U(t)$$

C_b^n is the attitude transformation matrix relating the body frame b with navigation frame n and is denoted as

$$C_b^n = \{C_{ij}\}_{i,j=1,2,3} \tag{12.3}$$

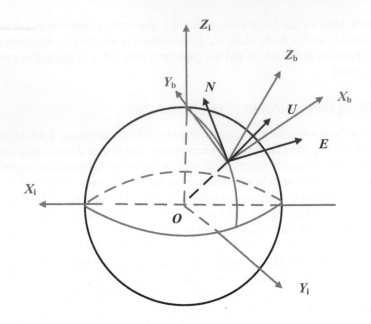

FIGURE 12.1   3D diagram of coordinate systems.

In frame n, the local gravity acceleration g and earth rate w_{ie} can be described by

$$g^n = [\; 0 \quad 0 \quad -g\;]^T \tag{12.4}$$

and

$$w_{ie}^n = [\; 0 \quad \Omega_N \quad \Omega_U\;]^T \tag{12.5}$$

The superscript n denotes the navigation frame. Ω_U and Ω_N are the up and north components of earth rate, respectively. The gyroscope is applied to measure the earth rate w_{ie} and the accelerometer is to measure local gravity acceleration g for INS on a ground stationary base.

Thus, the error equation with multiple disturbances can be further described as

$$\begin{bmatrix} \dot{\phi}_E(t) \\ \dot{\phi}_N(t) \\ \dot{\phi}_U(t) \end{bmatrix} = F_0 \begin{bmatrix} \phi_E(t) \\ \phi_N(t) \\ \phi_U(t) \end{bmatrix} + C_b^n \begin{bmatrix} \varepsilon_x(t) \\ \varepsilon_y(t) \\ \varepsilon_z(t) \end{bmatrix} + f_0(\phi) + B_1 d(t) \tag{12.6}$$

where the matrix F_0 is defined as

$$F_0 = \begin{bmatrix} 0 & \Omega_U & 0 \\ -\Omega_U & 0 & 0 \\ 0 & 0 & 0 \end{bmatrix} \tag{12.7}$$

and

$$f_0(\phi) = [\; -\Omega_N sin\phi_U \quad \Omega_N(1 - cos\phi_U) \quad -\Omega_N \overline{\phi_E}\;]^T \tag{12.8}$$

$d(t)$ represents the norm-bounded disturbance, which results from modeling error, system noise and uncertainties in the exogenous system. B_1 is a known coefficient matrix of the system with suitable dimensions. The subscripts E, N and U correspond to east, north, and up of the navigation frame n, and the subscript x, y and z denote right, front, up of the body frame b, respectively. Similarly to [21, 61], the level accelerometer outputs $f_E(t), f_N(t)$ and east gyro output $w_E(t)$ are selected as measured signals during the stationary base alignment. However, in this chapter we consider nonlinear dynamics and get the following measurement output equation

$$\begin{bmatrix} f_E(t) \\ f_N(t) \\ w_E(t) \end{bmatrix} = H \begin{bmatrix} \phi_E(t) \\ \phi_N(t) \\ \phi_U(t) \end{bmatrix} + D^b \omega^b(t) + g_0(\phi) + D_1 d(t) \tag{12.9}$$

where

$$H = \begin{bmatrix} 0 & g & 0 \\ -g & 0 & 0 \\ 0 & -\Omega_U & 0 \end{bmatrix} \tag{12.10}$$

$$D^b = \begin{bmatrix} 0 & 0 & 0 & C_{11} & C_{12} \\ 0 & 0 & 0 & C_{21} & C_{22} \\ C_{11} & C_{12} & C_{13} & 0 & 0 \end{bmatrix} \tag{12.11}$$

and

$$g_0(\phi) = \begin{bmatrix} 0 & 0 & \Omega_N sin\phi_U \end{bmatrix}^T \tag{12.12}$$

In (12.9), the inertial sensor bias error value is denoted by

$$\omega^b(t) = \begin{bmatrix} \varepsilon_x(t) & \varepsilon_y(t) & \varepsilon_z(t) & \nabla_x(t) & \nabla_y(t) \end{bmatrix}^T$$

where ∇ represents the accelerometer biases. The inertial sensor bias values can be supposed to satisfy first-order Gaussian Markov process, which can be described by

$$\dot{\omega}(t) = W\omega(t) + Ed(t) \tag{12.13}$$

where

$$W = diag\{-\frac{1}{\tau_1}, \cdots, -\frac{1}{\tau_5}\}$$

and τ_i $(i = 1, \cdots, 5)$ are correlation times of Markov process. In (12.9) and (12.13), H, D^b, D_1 and E represent the known coefficient matrices with suitable dimensions.

The relationships of gyro and accelerometer outputs between navigation frame and body frame can be denoted as

$$C_b^n \begin{bmatrix} \varepsilon_x & \varepsilon_y & \varepsilon_z \end{bmatrix}^T = \begin{bmatrix} \varepsilon_E & \varepsilon_N & \varepsilon_U \end{bmatrix}^T \tag{12.14}$$

and

$$C_b^n \begin{bmatrix} \nabla_x & \nabla_y & \nabla_z \end{bmatrix}^T = \begin{bmatrix} \nabla_E & \nabla_N & \nabla_U \end{bmatrix}^T \tag{12.15}$$

Based on (12.14) and (12.15), Equations (12.6) and (12.9) can be further formulated as

$$
\begin{cases}
\begin{bmatrix} \dot{\phi}_E(t) \\ \dot{\phi}_N(t) \\ \dot{\phi}_U(t) \end{bmatrix} = F_0 \begin{bmatrix} \phi_E(t) \\ \phi_N(t) \\ \phi_U(t) \end{bmatrix} + \begin{bmatrix} \varepsilon_E(t) \\ \varepsilon_N(t) \\ \varepsilon_U(t) \end{bmatrix} + f_0(\phi) + B_1 d(t) \\[20pt]
\begin{bmatrix} f_E(t) \\ f_N(t) \\ w_E(t) \end{bmatrix} = H \begin{bmatrix} \phi_E(t) \\ \phi_N(t) \\ \phi_U(t) \end{bmatrix} + \begin{bmatrix} \nabla_E(t) \\ \nabla_N(t) \\ \varepsilon_E(t) \end{bmatrix} + g_0(\phi) + D_1 d(t)
\end{cases}
\tag{12.16}
$$

Actually, the transformed system model (12.16) of SINS is similar to the model of GINS. From the definitions of the system matrix (12.7) and output matrix (12.10), it can be calculated that the observability index of pair (F_0, H) is equal to 2. The expansions of sinusoidal and cosine functions can be denoted as follows

$$
sin(\phi_U) = \phi_U - \frac{\phi_U^3}{6} + \frac{\phi_U^5}{120} + o(\phi_U^5),
\tag{12.17}
$$

and

$$
cos(\phi_U) = 1 - \frac{\phi_U^2}{2} + \frac{\phi_U^4}{24} + o(\phi_U^5).
\tag{12.18}
$$

Applying (12.17) and (12.18) to the nonlinear terms of the error model (12.16), the nonlinear INS error system (12.16) can be formulated by

$$
\begin{cases}
\dot{x}(t) = Ax(t) + Ff(x(t)) + B\omega(t) + u(t) + B_1 d(t) \\
y(t) = Cx(t) + Gg(x(t)) + D\omega(t) + u(t) + D_1 d(t)
\end{cases}
\tag{12.19}
$$

where the system state

$$
x^T(t) = \begin{bmatrix} \phi_E(t) & \phi_N(t) & \phi_U(t) \end{bmatrix}
$$

and the system output variable

$$
y^T(t) = \begin{bmatrix} f_E(t) & f_N(t) & w_E(t) \end{bmatrix}
$$

$u(t)$ is the control input for calibrating and compensating.

$$
\omega(t) = \begin{bmatrix} \varepsilon_E(t) & \varepsilon_N(t) & \varepsilon_U(t) & \nabla_E(t) & \nabla_N(t) \end{bmatrix}^T
$$

is the inertial sensor bias values in frame n. B_1 and D_1 are denoted in (12.6) and (12.9), respectively. And,

$$
A = \begin{bmatrix} 0 & \Omega_U & -\Omega_N \\ -\Omega_U & 0 & 0 \\ \Omega_N & 0 & 0 \end{bmatrix}, \quad C = \begin{bmatrix} 0 & g & 0 \\ -g & 0 & 0 \\ 0 & -\Omega_U & \Omega_N \end{bmatrix},
$$

$$B = \begin{bmatrix} 1 & 0 & 0 & 0 & 0 \\ 0 & 1 & 0 & 0 & 0 \\ 0 & 0 & 1 & 0 & 0 \end{bmatrix}, D = \begin{bmatrix} 0 & 0 & 1 & 0 & 0 \\ 0 & 0 & 0 & 1 & 0 \\ 1 & 0 & 0 & 0 & 0 \end{bmatrix}.$$

The nonlinear terms can be denoted as

$$f(x(t)) = \begin{bmatrix} -\frac{\phi_U^3}{6} + \frac{\phi_U^5}{120} & \frac{\phi_U^2}{2} - \frac{\phi_U^4}{24} & \overline{\overline{\phi_E}} \end{bmatrix}^T, \tag{12.20}$$

and

$$g(x(t)) = \begin{bmatrix} 0 & 0 & -\frac{\phi_U^3}{6} + \frac{\phi_U^5}{120} \end{bmatrix}^T \tag{12.21}$$

where,

$$\overline{\overline{\phi_E}} = \phi_N(\phi_U - \frac{\phi_U^3}{6} + \frac{\phi_U^5}{120}) - \phi_E(-\frac{\phi_U^2}{2} + \frac{\phi_U^4}{24})$$

Correspondingly, the parameter matrices are denoted by

$$F = diag\{-\Omega_N, \Omega_N, -\Omega_N\}, G = diag\{0, 0, \Omega_N\}.$$

$d(t)$ is the norm-bounded disturbance. For the stationary base initial alignment, the initial misalignment angles should be in a proper interval. Therefore, the high-order infinitesimal term $o(\phi_U^5)$ can be merged into the norm-bounded disturbance $d(t)$.

After the INS error model is established, the nonlinear initial alignment for INS on a ground stationary base can be transformed into a robust filter design problem for nonlinear systems with multiple disturbances. If $F = 0$ and $G = 0$, the nonlinear model (12.19) can be degenerated into a linear model. The nonlinear dynamics existing in both the state equation and the output equation of the dynamical systems forms the key difficulty in the following nonlinear robust filter design procedures.

It is noted that $f(0) = 0$ and $g(0) = 0$ hold. In order to provide feasible algorithms, we adopt the following assumptions so that nonlinear robust filtering can be applied. $f(x(t))$ is supposed to satisfy

$$\|f(x_1(t)) - f(x_2(t))\| \leq \|U_1(x_1(t) - x_2(t))\| \tag{12.22}$$

for any $x_1(t)$ and $x_2(t)$, and for a known matrix U_1. $g(x(t))$ satisfies

$$\|g(x_1(t)) - g(x_2(t))\| \leq \|U_2(x_1(t) - x_2(t))\| \tag{12.23}$$

for any $x_1(t)$ and $x_2(t)$, and for a known matrix U_2. In this chapter, the constant matrices U_1 and U_2 are selected as

$$U_1 = \begin{bmatrix} 0 & 0 & \frac{\pi^2}{162} \\ 0 & 0 & \frac{\pi}{9} \\ \frac{\pi^2}{162} & \frac{\pi}{9} & \frac{9\pi + \pi^2}{81} \end{bmatrix}, U_2 = \begin{bmatrix} 0 & 0 & 0 \\ 0 & 0 & 0 \\ 0 & 0 & \frac{\pi^2}{162} \end{bmatrix} \tag{12.24}$$

Up to now, a new nonlinear INS model has been established, for which nonlinear robust filtering can be provided for initial alignment problems.

12.3　ROBUST ANTI-DISTURBANCE FILTER DESIGN

In this section, the intention is to design a robust anti-disturbance filter in the presence of multiple disturbances for the nonlinear systems concerned. Drift estimations will be applied to reject the inertial sensor drift. H_∞ optimization technique will be adopted to attenuate the norm-bounded disturbances and guarantee cost performance to optimize the error variance. Accordingly, the proposed multi-objective filter has both disturbance rejection and attenuation performance. See the block diagram of the robust anti-disturbance filter design in Figure 12.2.

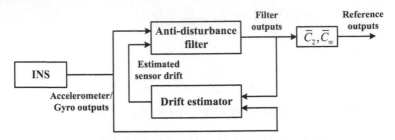

FIGURE 12.2　The block diagram of robust anti-disturbance filter design.

For this purpose, we firstly construct the inertial sensor drift estimator as follows

$$\dot{\widehat{\omega}}(t) = W\widehat{\omega}(t) + K[y(t) - \widehat{y}(t)] \tag{12.25}$$

$\widehat{\omega}(t)$ is the estimation of inertial sensor drifts $\omega(t)$. Based on the (12.25), the mixed multi-objective filter is constructed

$$\begin{cases} \dot{\widehat{x}}(t) &=& A\widehat{x}(t) + Ff(\widehat{x}(t)) + u(t) + u_{c1}(t) + L[y(t) - \widehat{y}(t)] \\ \widehat{y}(t) &=& C\widehat{x}(t) + Gg(\widehat{x}(t)) + u(t) + u_{c2}(t) \end{cases} \tag{12.26}$$

where $\widehat{x}(t)$ is the estimation of state $x(t)$, $\widehat{y}(t)$ is the estimation of output signal $y(t)$. We use $u_{c1}(t) = B\widehat{\omega}(t)$ and $u_{c2}(t) = D\widehat{\omega}(t)$ as the compensation terms, which are applied to reject inertial sensor biases. K and L are the filter gains to be determined.

Denoting

$$\widetilde{x}(t) = x(t) - \widehat{x}(t), \widetilde{\omega}(t) = \omega(t) - \widehat{\omega}(t),$$

$$\widetilde{f}(t) = f(x(t)) - f(\widehat{x}(t)), \widetilde{g}(t) = g(x(t)) - g(\widehat{x}(t)),$$

the estimation error system yields

$$\begin{cases} \dot{\widetilde{x}}(t) &=& (A - LC)\widetilde{x}(t) + F\widetilde{f}((t)) + (B - LD)\widetilde{\omega^n}(t) \\ && -LG\widetilde{g}(t) + (B_1 - LD_1)d(t) \\ \dot{\widetilde{\omega}}(t) &=& (W - KD)\widetilde{\omega}(t) - KC\widetilde{x}(t) - KG\widetilde{g}(t) + (E - KD_1)d(t) \end{cases} \tag{12.27}$$

For Equation (12.27), the next step is to consider its mixed multi-objective performance indices. Combining the reference outputs with estimation error Equation (12.27) yields

$$\begin{cases} \begin{bmatrix} \dot{\widetilde{x}}(t) \\ \dot{\widetilde{\omega}}(t) \end{bmatrix} = \overline{A} \begin{bmatrix} \widetilde{x}(t) \\ \widetilde{\omega}(t) \end{bmatrix} + \begin{bmatrix} F\widetilde{f}(t) \\ 0 \end{bmatrix} + \overline{B}d(t) - \begin{bmatrix} LG \\ KG \end{bmatrix} \widetilde{g}(t) \\[4mm] z_\infty(t) = \overline{C_\infty} \begin{bmatrix} \widetilde{x}(t) \\ \widetilde{\omega}(t) \end{bmatrix} \\[4mm] z_2(t) = \overline{C_2} \begin{bmatrix} \widetilde{x}(t) \\ \widetilde{\omega}(t) \end{bmatrix} \end{cases} \tag{12.28}$$

where

$$\overline{A} = \begin{bmatrix} A - LC & B - LD \\ -KC & W - KD \end{bmatrix}, \overline{B} = \begin{bmatrix} B_1 - LD_1 \\ E - KD_1 \end{bmatrix}$$

$$\overline{C_2} = [\ C_{21} \quad C_{22}\ ], \overline{C_\infty} = [\ C_{\infty 1} \quad C_{\infty 2}\ ].$$

$z_\infty(t)$, $z_2(t)$ are H_∞ and guarantee cost performance outputs, respectively. C_{21}, C_{22}, $C_{\infty 1}$, $C_{\infty 2}$ are selected weighting matrices. The following definitions can be seen in [95].

Definition 12.1 The generalized H_∞ performance measurement for (12.28) is defined as

$$J_\infty = \|\ z_\infty(t)\ \|^2 - \gamma \|\ d(t)\ \|^2 - \phi^T(0)P\phi(0) \tag{12.29}$$

where $\gamma > 0$ is scalar and P is a positive definite matrix.

Definition 12.2 The guarantee cost performance measurement for Equation (12.28) is defined as

$$J_2 = \|\ z_2(t)\ \|^2$$

At this stage, the objective is to find K and L such that system (12.28) is stable and satisfies the mixed performance indices. The following result provides a new robust anti-disturbance filter design method for initial alignment of INS based on linear matrix inequalities (LMIs).

Theorem 12.1

For the weighting constants $r_1 > 0, r_2 > 0, \lambda_1 > 0, \lambda_2 > 0$, tuning parameters $\alpha > 0$, $\alpha_1 > 0$ and matrices $C_{21}, C_{22}, C_{\infty 1}, C_{\infty 2}$, if the following LMI-based optimization problem:

$$min\{r_1\gamma + r_2\phi^T(0)P_1\phi(0)\}$$

subject to

$$
\begin{bmatrix}
\Xi_1 & \Xi_{12} & \Xi_{13} & \Xi_{14} & P_2F & \Xi_{16} & \Xi_{17} & C_{\infty 1}^T & C_{21}^T \\
* & \Xi_2 & \Xi_{23} & 0 & \alpha P_2F & \Xi_{26} & \Xi_{27} & 0 & 0 \\
* & * & \Xi_3 & \Xi_{34} & 0 & \Xi_{36} & \Xi_{37} & C_{\infty 2}^T & C_{22}^T \\
* & * & * & \Xi_4 & 0 & \Xi_{46} & \Xi_{47} & 0 & 0 \\
* & * & * & * & -\lambda_1 I & 0 & 0 & 0 & 0 \\
* & * & * & * & * & -\lambda_2 I & 0 & 0 & 0 \\
* & * & * & * & * & * & -\gamma I & 0 & 0 \\
* & * & * & * & * & * & * & -I & 0 \\
* & * & * & * & * & * & * & * & -I
\end{bmatrix} < 0 \qquad (12.30)
$$

where

$$\Xi_1 = sym(P_2A - R_1C) + \lambda_1 U_1^T U_1 + \lambda_2 U_2^T U_2,$$

$$\Xi_{12} = P_1 - P_2 + \alpha(P_2A - R_1C)^T,$$

$$\Xi_{13} = P_2B - R_1D - C^T R_2^T, \quad \Xi_{14} = -\alpha_1 C^T R_2^T,$$

$$\Xi_{16} = -R_1G, \quad \Xi_{17} = P_2B_1 - R_1D_1,$$

$$\Xi_2 = -\alpha P_2 - \alpha P_2^T, \quad \Xi_{23} = \alpha(P_2B - R_1D),$$

$$\Xi_{26} = -\alpha R_1G, \quad \Xi_{27} = \alpha(P_2B_1 - R_1D_1),$$

$$\Xi_3 = sym(P_4W - R_2D), \quad \Xi_4 = -\alpha_1 P_4 - \alpha_1 P_4^T$$

$$\Xi_{34} = P_3 - P_4 + \alpha_1(P_4W - R_2D)^T,$$

$$\Xi_{36} = -R_2G, \quad \Xi_{37} = P_4E - R_2D_1,$$

$$\Xi_{46} = -\alpha_1 R_2G, \quad \Xi_{47} = \alpha_1(P_4E - R_2D_1).$$

are feasible with respect to matrices $P_i > 0$ ($i = 1, 3$), P_2, P_4 and R_1, R_2, then there exists a mixed multi-objective filter with gains $L = P_2^{-1}R_1$ and $K = P_4^{-1}R_2$ such that the error system (12.28) is stable, $J_2 \leq x^T(0)P_1x(0)$ and $J_\infty < 0$. ∎

Proof. Let

$$
\begin{aligned}
V_1(t) &= \tilde{x}^T(t)P_1\tilde{x}(t) + \lambda_1 \int_0^t [\|U_1\tilde{x}(\tau)\|^2 - \|\tilde{f}(\tau)\|^2]d\tau \\
&\quad + \lambda_2 \int_0^t [\|U_2\tilde{x}(\tau)\|^2 - \|\tilde{g}(\tau)\|^2]d\tau \\
&= \begin{bmatrix} \tilde{x}(t) \\ \dot{\tilde{x}}(t) \end{bmatrix}^T \begin{bmatrix} I & 0 \\ 0 & 0 \end{bmatrix} \begin{bmatrix} P_1 & 0 \\ P_2^T & \alpha P_2^T \end{bmatrix} \begin{bmatrix} \tilde{x}(t) \\ \dot{\tilde{x}}(t) \end{bmatrix} \\
&\quad + \lambda_1 \int_0^t [\|U_1\tilde{x}(\tau)\|^2 - \|\tilde{f}(\tau)\|^2]d\tau \\
&\quad + \lambda_2 \int_0^t [\|U_2\tilde{x}(\tau)\|^2 - \|\tilde{g}(\tau)\|^2]d\tau
\end{aligned}
$$

$$
\begin{aligned}
V_2(t) &= \tilde{\omega}^T(t)P_3\tilde{\omega}(t) \\
&= \begin{bmatrix} \tilde{\omega}(t) \\ \dot{\tilde{\omega}}(t) \end{bmatrix}^T \begin{bmatrix} I & 0 \\ 0 & 0 \end{bmatrix} \begin{bmatrix} P_3 & 0 \\ P_4^T & \alpha_1 P_4^T \end{bmatrix} \begin{bmatrix} \tilde{\omega}(t) \\ \dot{\tilde{\omega}}(t) \end{bmatrix}
\end{aligned}
$$

Consider the following Lyapunov function:

$$
V(t) = V_1(t) + V_2(t) \tag{12.31}
$$

It is verified that $V(t) \geq 0$ holds for all arguments. Variable t will be omitted in the following procedures for simplification, if there are no various interpretations. Along with the trajectories of (12.27), it can be shown that

$$
\begin{aligned}
\dot{V}_1 &= 2\tilde{x}^T P_1\dot{\tilde{x}} + \lambda_1[\|U_1\tilde{x}\|^2 - \|\tilde{f}\|^2] + \lambda_2[\|U_2\tilde{x}\|^2 - \|\tilde{g}\|^2] \\
&= 2\begin{bmatrix} \tilde{x} \\ \dot{\tilde{x}} \end{bmatrix}^T \begin{bmatrix} P_1 & P_2 \\ 0 & \alpha P_2 \end{bmatrix} \begin{bmatrix} \dot{\tilde{x}} \\ 0 \end{bmatrix} + \lambda_1[\|U_1\tilde{x}\|^2 - \|\tilde{f}\|^2] \\
&\quad + \lambda_2[\|U_2\tilde{x}\|^2 - \|\tilde{g}\|^2] \\
&= 2\tilde{x}^T P_1\dot{\tilde{x}} - 2\tilde{x}^T P_2\dot{\tilde{x}} + 2\tilde{x}^T P_2(B - LD)\tilde{\omega} + 2\tilde{x}^T P_2(A - LC)\tilde{x} \\
&\quad + 2\dot{\tilde{x}}^T \alpha P_2(A - LC)\tilde{x} - 2\dot{\tilde{x}}^T \alpha P_2\dot{\tilde{x}} + 2\tilde{x}^T P_2(B_1 - LD_1)d \\
&\quad + 2\tilde{x}^T P_2\tilde{f} + 2\dot{\tilde{x}}^T \alpha P_2(B_1 - LD_1)d + 2\dot{\tilde{x}}^T \alpha P_2(B - LD)\tilde{\omega} \\
&\quad + 2\dot{\tilde{x}}^T \alpha P_2\tilde{f} + \lambda_1\tilde{x}^T U_1^T U_1\tilde{x} + \lambda_2\tilde{x}^T U_2^T U_2\tilde{x} - 2\tilde{x}^T P_2 LG\tilde{g} \\
&\quad - 2\dot{\tilde{x}}^T \alpha P_2 LG\tilde{g} - \lambda_1\|\tilde{f}\|^2 - \lambda_2\|\tilde{g}\|^2
\end{aligned}
$$

$$\dot{V}_2 = 2\tilde{\omega}^T P_3 \dot{\tilde{\omega}}$$

$$= 2\begin{bmatrix} \tilde{\omega} \\ \dot{\tilde{\omega}} \end{bmatrix}^T \begin{bmatrix} P_3 & P_4 \\ 0 & \alpha_1 P_4 \end{bmatrix} \begin{bmatrix} \dot{\tilde{\omega}} \\ 0 \end{bmatrix}$$

$$= 2\tilde{\omega}^T P_3 \dot{\tilde{\omega}} - 2\tilde{\omega}^T P_4 \dot{\tilde{\omega}} + 2\dot{\tilde{\omega}}^T \alpha_1 P_4 (W - KD)\tilde{\omega} - 2\dot{\tilde{\omega}}^T \alpha_1 P_4 \dot{\tilde{\omega}}$$

$$+ 2\tilde{\omega}^T P_4 (W - KD)\tilde{\omega} + 2\tilde{\omega}^T P_4 (E - KD_1)d$$

$$+ 2\dot{\tilde{\omega}}^T \alpha_1 P_4 (E - KD_1)d - 2\tilde{\omega}^T P_4 KC\tilde{x} - 2\dot{\tilde{\omega}}^T \alpha_1 P_4 KC\tilde{x}$$

$$- 2\tilde{\omega}^T P_4 KG\tilde{g} - 2\dot{\tilde{\omega}}^T \alpha_1 P_4 KG\tilde{g}$$

In the absence of $d(t)$ (i.e., $d(t) = 0$), it can be seen that

$$\dot{V}(t) = s^T(t)[\Phi - \zeta(t)\zeta^T(t)]s(t)$$

where

$$\Phi = \begin{bmatrix} \Phi_{11} & \Xi_{12} & \Xi_{13} & \Xi_{14} & P_2 F & \Xi_{16} \\ * & \Xi_2 & \Xi_{23} & 0 & \alpha P_2 F & \Xi_{26} \\ * & * & \Phi_{33} & \Xi_{34} & 0 & \Xi_{36} \\ * & * & * & \Xi_4 & 0 & \Xi_{46} \\ * & * & * & * & -\lambda_1 I & 0 \\ * & * & * & * & * & -\lambda_2 I \end{bmatrix}$$

$$s^T(t) = \begin{bmatrix} \tilde{x}^T & \dot{\tilde{x}}^T & \tilde{\omega}^T & \dot{\tilde{\omega}}^T & \tilde{f}^T & \tilde{g}^T \end{bmatrix}$$

$$\zeta^T(t) = \begin{bmatrix} C_{21} & 0 & C_{22} & 0 & 0 & 0 \end{bmatrix}$$

$$\Phi_{11} = \Xi_1 + C_{21}^T C_{21}, \Phi_{33} = \Xi_3 + C_{22}^T C_{22}.$$

It can be seen by using the Schur complement formula that (12.30) leads to $\Phi < 0$. When $\Phi < 0$ holds, we have

$$\dot{V}(t) < -s^T(t)\zeta(t)\zeta^T(t)s(t) = -z_2^T z_2 \leq 0$$

Consider two auxiliary functions as follows

$$J_0 = z_2^T(t)z_2(t) + \dot{V}(t),$$

$$J_1 = z_\infty^T(t)z_\infty(t) - \gamma d(t)^T d(t) + \dot{V}(t)$$

In the presence of $d(t)$, it can be concluded that

$$J_1 = s_1^T(t)\Psi s_1(t),$$

where $s_1^T(t) = \begin{bmatrix} s^T(t) & d^T(t) \end{bmatrix}$, and

$$\Psi = \begin{bmatrix} \Xi_1 + C_{\infty1}^T C_{\infty1} & \Xi_{12} & \Xi_{13} + C_{\infty1}^T C_{\infty2} & \Xi_{14} & P_2 F & \Xi_{16} & \Xi_{17} \\ * & \Xi_2 & \Xi_{23} & 0 & \alpha P_2 F & \Xi_{26} & \Xi_{27} \\ * & * & \Xi_3 + C_{\infty2}^T C_{\infty2} & \Xi_{34} & 0 & \Xi_{36} & \Xi_{37} \\ * & * & * & \Xi_4 & 0 & \Xi_{46} & \Xi_{47} \\ * & * & * & * & -\lambda_1 I & 0 & 0 \\ * & * & * & * & * & -\lambda_2 I & 0 \\ * & * & * & * & * & * & -\gamma I \end{bmatrix}$$

Using the Schur complement formula again to (12.30) yields $\Psi < 0$. Therefore, it can be claimed that both $J_0 < 0$ and $J_1 < 0$ can be guaranteed under the condition (12.30). Integrating J_0 and J_1 from initial time to ∞ implies that $J_2 \leq x^T(0)P_1x(0)$ and $J_\infty < 0$. This completes the proof. $\qquad\square$

Theorem 12.1 provides a robust initial alignment design method based on the mixed multi-objective optimization principle. It also differs from the main results in [21, 71, 95], since it is based on the augmented Lyapunov functional approach ([134]). This result involves the tuning parameters α, α_1 and slack variables P_2 and P_4, and may lead to less conservative solutions.

12.4 SIMULATION EXAMPLES

In this section, SINS on a ground stationary base are considered with multiple disturbances. We consider the SINS data obtained in an experiment in the laboratory.

Simulation conditions:

1. Local latitude is $39.9deg$ north latitude;
2. Initial misalignment angles ϕ_U, ϕ_E, ϕ_N are chosen as 5, 0.5 and $0.5deg$, respectively;
3. Gyroscopes: the relative times of Markov process are $\tau_1 = \tau_2 = \tau_3 = 3600s$ and random bias is $0.05deg/hr$;
4. Accelerometers: the relative times of Markov process are $\tau_4 = \tau_5 = 1800s$ and random bias is $50\mu g$.

The correlative times for the random bias are decided based on the properties of the selected inertial sensors by testing. The coefficient matrices of nonlinear model (12.19) are given by

$$A = 10^{-4} \times \begin{bmatrix} 0 & 0.467752 & -0.559426 \\ -0.467752 & 0 & 0 \\ 0.559426 & 0 & 0 \end{bmatrix},$$

$$C = \begin{bmatrix} 0 & 9.801609 & 0 \\ -9.801609 & 0 & 0 \\ 0 & -0.000047 & 0.000056 \end{bmatrix},$$

$$E = \left[\begin{array}{ccccc} \frac{0.0001\pi}{648000} & \frac{0.0001\pi}{648000} & \frac{0.0001\pi}{648000} & 10^{-7}g & 10^{-7}g \end{array}\right]^T,$$

$$B_1 = \left[\begin{array}{ccc} \frac{0.01\pi}{648000} & \frac{0.01\pi}{648000} & \frac{0.1\pi}{648000} \end{array}\right]^T,$$

$$D_1 = \left[\begin{array}{ccc} 10^{-5}g & 10^{-5}g\frac{0.1\pi}{648000} \end{array}\right]^T,$$

$$F = 0.559426 \times 10^{-4} diag\{-1,1,-1\},$$

$$G = 0.559426 \times 10^{-4} diag\{0,0,1\}.$$

For an outdoor SINS, machine vibrations, modeling errors and outdoor environmental disturbances unavoidably exist. These unknown and uncertain variables can be merged into $d(t)$ as a norm-bounded disturbance for the SINS, where

$$d^T(t) = \left[\begin{array}{ccc} 0.001sign(sin(100t)) & 0.001sign(sin(100t)) & 0.0001sign(sin(100t)) \end{array}\right].$$

We insert the non-Gaussian disturbance $d(t)$ into the SINS data measured in the laboratory. The H_∞ reference output weighting matrices are selected as

$$C_{\infty 1} = \left[\begin{array}{ccc} 0.1 & 0.1 & 0.01 \end{array}\right], C_{\infty 2} = \left[\begin{array}{ccccc} 0 & 0 & 0 & 0 & 0 \end{array}\right].$$

The H_2 reference output weighting matrices are selected as

$$C_{21} = \left[\begin{array}{ccc} 10^{-6}g & 10^{-6}g & 10^{-5}g \end{array}\right],$$

$$C_{22} = \left[\begin{array}{ccccc} \frac{0.01\pi}{180} & \frac{0.01\pi}{180} & \frac{0.01\pi}{180} & \frac{0.1\pi}{180} & \frac{0.1\pi}{180} \end{array}\right].$$

For weighting parameters $r_1 = 1$, $r_2 = 10$, $\lambda_1 = 0.013211$, $\lambda_2 = 0.013211$, tuning parameters $\alpha = 12$ and $\alpha_1 = 12$, it can be solved via LMI related to (12.30) that $\gamma = 2.179985 \times 10^{-11}$, and the gain of robust multi-objective filter (12.26) is

$$L = \left[\begin{array}{ccc} 0.003684 & 0.012166 & -0.226204 \\ -0.015101 & -0.006456 & 0.547060 \\ -0.008493 & -0.019773 & 1.583700 \end{array}\right]$$

the gain of drift estimator is

$$K = 0.01 \times \left[\begin{array}{ccc} 0.000209 & 0.000208 & 0.041380 \\ 0.000250 & 0.000248 & 0.039678 \\ 0.000240 & 0.000237 & 0.040216 \\ 0.046968 & 0.046520 & 8.389645 \\ 0.046969 & 0.046520 & 8.389651 \end{array}\right].$$

TABLE 12.1

The STD of estimation error of misalignment angles ($10^{-5}rad$) for nonlinear model.

Method	leveling x	leveling y	azimuth
In this chapter	4.084436	5.962061	8.382810
UKF	27.573893	75.867469	18.481666

The parameter matrices of unscented KF are selected as follows. The initial mean square error matrix is

$$P(0) = diag\{(1deg)^2, (1deg)^2, (10deg)^2, (0.2deg/hr)^2,$$
$$(0.2deg/hr)^2, (0.2deg/hr)^2, (200\mu g)^2, (200\mu g)^2\}$$

The variance matrix of system noise is

$$Q = diag\{(0.1deg/hr)^2, (0.1deg/hr)^2, (0.1deg/hr)^2\},$$

and variance matrix of measurement noise is selected as

$$R = diag\{(50\mu g)^2, (50\mu g)^2, (0.05deg/h)^2\}.$$

The estimation errors of three misalignment angles are described in Figures 12.3-12.5, where solid lines represent the estimation errors of the robust anti-disturbance filter in this chapter and dash-dot lines denote the estimation errors based on UKF. For nonlinear systems with multiple disturbances, the SINS is mainly affected by exogenous disturbance, measurement noise and inertial sensor random biases. From Figures 12.3-12.5, it can be seen that the proposed robust anti-disturbance filter has a good ability for initial alignment of SINS with multiple disturbances, and the UKF is degraded when the concerned SINS has non-Gaussian noise. The composition of standard deviations (STD) of three misalignment angle estimation errors (after the 30th second) between the robust anti-disturbance filter in this chapter and UKF is shown in Table 12.1. In Table 12.1, it is shown that the accuracy of the proposed robust filter is higher than UKF for the considered systems with multiple disturbances.

When three misalignment angles are selected as 1, 0.1, 0.1 deg, the azimuth angle estimation error is seen in Figure 12.6, where solid lines represent the estimation errors based on the nonlinear model in this chapter and dash-dot lines denote the estimation errors based on the linear model in [21]. The standard deviations of three misalignment angle estimation errors (after the 30th second) between the nonlinear model in this chapter and the linear model is shown in Table 12.2. From Figure 12.6 and Table 12.2, we can see that the accuracy of the nonlinear model is higher than the linear model, where the linearization of heading error leads to modeling error.

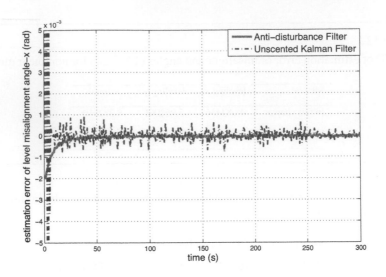

FIGURE 12.3   Estimation errors of ϕ_x for nonlinear model.

TABLE 12.2

The STD of estimation error of misalignment angles ($10^{-4}rad$).

Method	leveling x	leveling y	azimuth
Nonlinear model	0.032142	0.042950	2.103586
Linear model	0.032335	0.047724	3.122321

12.5   CONCLUSION

In this chapter, the problem of nonlinear initial alignment is investigated for INS with multiple disturbances. A new INS nonlinear error model for a stationary base is established, and is transformed into a robust filter design problem for systems with multiple disturbances. The multiple disturbances are supposed to include inertial sensor drifts and norm-bounded disturbances. As such, we present a new anti-disturbance filtering method, where drift estimations are applied to reject inertial sensor random biases, H_∞ optimization technique is adopted to attenuate norm-bounded uncertain disturbances and guarantee cost performance to optimize the error variance. Accordingly, the proposed anti-disturbance filter has enhanced disturbance rejection and attenuation performance.

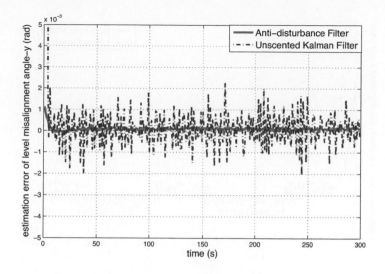

FIGURE 12.4   Estimation errors of ϕ_y for nonlinear model.

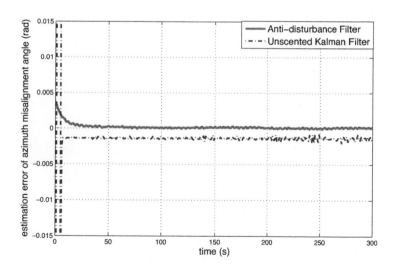

FIGURE 12.5   Estimation errors of ϕ_z for nonlinear model.

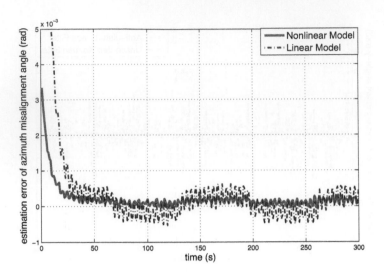

FIGURE 12.6   The comparison of estimation error of ϕ_z.

13 Robust Fault Diagnosis Method Based on Disturbance Observer

13.1 INTRODUCTION

Automatic control systems are becoming more and more sophisticated and the control algorithms more and more complex. To improve the reliability of control systems, much attention has been paid to the problem of fault detection and diagnosis (FDD) in dynamic systems over the past two decades. Fruitful results can be found in references [8, 76, 109]. Effective methodologies mainly include observer (or filter)-based schemes, identification-based approaches, and statistic methods. Among these methods, observer-based approaches have been effectively applied to dynamic systems, where observer signals are used to infer information about the fault. Many significant approaches have been presented and applied to practical processes successfully (see [34, 71, 90, 115, 168, 229], and references therein).

External disturbances and unmodeled dynamics as well as parametric variations exist widely in practical processes. The situation will be much more complicated when faults and disturbances exist simultaneously in FDD problems. To overcome these obstacles, some elegant approaches have been provided and applied in practical engineering (see [26, 52, 56, 73, 92, 96]). An optimal fault diagnosis observer was proposed in [26], which can produce disturbance decoupled state estimation with minimum variance for linear time varying Gaussian systems with both noise and unknown disturbances. Sliding mode observers were designed to reconstruct the disturbance inputs and then to generate an effective residual [56]. This approach can be invalid when the fundamental problem of residual generation is unsolvable. In [73], a modified proportional, derivative and integral observer was developed to estimate both states, input disturbances and measurement noises, where the disturbances were supposed to be constants. In [92], an entropy optimization principle was applied to maximize the entropies when faults occur and to minimize the entropies resulting from other stochastic disturbances. H_∞ optimization was applied for fault detection in [52]. In [96], a generalized H_∞ technique was used to attenuate disturbances for fault diagnosis problems. However, in [52, 92] the authors only focused on fault detection problems.

The idea of disturbance attenuation was applied in most of the presented results for fault diagnosis problems. To improve the accuracy of diagnosis, the idea of disturbance rejection should be discussed for fault diagnosis problems. Disturbance observer based control (DOBC) strategies appeared in the late 1980s and have been applied in many control areas (see [87], and references therein). It has been shown that in many applications DOBC is a robust control scheme against various unknown disturbances. In this chapter, the disturbances represented by a linear exogenous system, not restricted to being constant, harmonic or neutral stable, are estimated by the disturbance observer. A composite observer is designed combining the fault diagnosis observer and disturbance observer, with which the fault can be diagnosed with disturbance rejection performance.

13.2   PROBLEM FORMULATION AND PRELIMINARIES

Consider the following nonlinear system with faults and external disturbances

$$\dot{x}(t) = Ax(t) + Gg(x(t)) + Hu(t) + JF(t) + Hd(t) \tag{13.1}$$

where $x(t) \in R^m$ is the measurable state, $u(t)$ is the control input. $F(t)$ is the fault vector to be diagnosed, and is supposed to occur at time $T > 0$ as a constant. $d(t)$ is the unknown disturbance and A, G, H, J represent the coefficient matrices of the weight system with suitable dimensions. For a known matrix U_1, $g(x(t))$ is a nonlinear vector function that is supposed to satisfy $g(0) = 0$ and the following norm condition

$$\|g(x_1(t)) - g(x_2(t))\| \le \|U_1[x_1(t) - x_2(t)]\| \tag{13.2}$$

for any $x_1(t)$ and $x_2(t)$. The unknown external disturbance $d(t)$ is supposed to be generated by a linear exogenous system described by

$$\begin{cases} \dot{\omega}(t) & = & W\omega(t) \\ d(t) & = & E\omega(t) \end{cases} \tag{13.3}$$

where $\omega(t) \in R^p$ is the state variable, W, E are the known parameter matrices of the exogenous system.

Including unknown constant or harmonics with unknown phase and magnitude, this model can describe many kinds of disturbances in practical processes which are produced by periodic disturbances in vibrating structures, or rotating mechanisms with eccentricity [87]. The following assumption is required so that the concerned problem can be well-posed in this chapter.

Assumption 13.1 (W, HE) *is observable.*

13.3   FAULT DIAGNOSIS BASED ON DISTURBANCE OBSERVER

When all states of a system are available, it is unnecessary to estimate the states, and only the estimation of the disturbances needs to be undertaken [85]. In order to reject the disturbances, the disturbance observer should be constructed.

$$\begin{cases} \hat{d}(t) & = & E\hat{\omega}(t) \\ \hat{\omega}(t) & = & r(t) - Lx(t) \end{cases} \tag{13.4}$$

where $\hat{d}(t)$ is the estimation of external disturbance $d(t)$, $\hat{\omega}(t)$ is the estimation of $\omega(t)$, matrix L is the disturbance observer gain to be determined later and $r(t)$ is the auxiliary variable generated by

$$
\begin{aligned}
\dot{r}(t) \ = \ & (W + LHE)[r(t) - Lx(t)] \\
& -L[-Ax(t) - Hu(t) - Gg(x(t))]
\end{aligned}
\tag{13.5}
$$

By defining $e_\omega(t) = \omega(t) - \hat{\omega}(t)$, the disturbance estimation error system can be obtained from (13.1), (13.3) and (13.4-13.5) to show

$$
\dot{e}_\omega(t) = (W + LHE)e_\omega(t) + LJF(t)
\tag{13.6}
$$

Fault diagnosis needs to be accomplished in order to estimate the size of the fault. For this purpose, the following fault diagnosis observer with regard to (13.1) is constructed to reject the disturbance and estimate the fault.

$$
\begin{cases}
\dot{\hat{x}}(t) \ = \ & A\hat{x}(t) + Gg(\hat{x}(t)) + Hu(t) + J\hat{F}(t) \\
& +K[x(t) - \hat{x}(t)] + H\hat{d}(t) \\
\dot{\hat{F}}(t) \ = \ & -\Gamma_1\hat{F} + \Gamma_2[x(t) - \hat{x}(t)]
\end{cases}
\tag{13.7}
$$

where $\hat{x}(t)$ is the estimation of $x(t)$ in fault diagnosis observer, $\hat{F}(t)$ is the estimation of $F(t)$. K is the fault diagnosis observer gain to be determined. Γ_i $(i = 1, 2)$ $(\Gamma_1 > 0)$ are two operators also to be determined later. $\hat{d}(t)$ is the estimation of disturbance in (13.4), which is applied to compensate for the disturbance. Denoting

$$
\widetilde{F}(t) = F(t) - \hat{F}(t), e_x(t) = x(t) - \hat{x}(t),
$$

the estimation error system yields

$$
\begin{aligned}
\dot{e}_x(t) \ = \ & (A - K)e_x(t) + G[g(x(t)) - g(\hat{x}(t))] \\
& +J\widetilde{F}(t) + H[d(t) - \hat{d}(t)]
\end{aligned}
\tag{13.8}
$$

$$
\begin{aligned}
\dot{\widetilde{F}}(t) \ = \ & \Gamma_1\hat{F}(t) - \Gamma_1 F(t) - \Gamma_2 e_x(t) + \Gamma_1 F(t) \\
= \ & -\Gamma_1\widetilde{F}(t) - \Gamma_2 e_x(t) + \Gamma_1 F(t)
\end{aligned}
\tag{13.9}
$$

Combining estimation error Equation (13.6) with Equations (13.8-13.9) yields

$$
\begin{bmatrix} \dot{e}_\omega(t) \\ \dot{e}_x(t) \\ \dot{\widetilde{F}}(t) \end{bmatrix} = \bar{A} \begin{bmatrix} e_\omega(t) \\ e_x(t) \\ \widetilde{F}(t) \end{bmatrix} + \begin{bmatrix} LJ \\ 0 \\ \Gamma_1 \end{bmatrix} F(t)
$$

$$
+ \begin{bmatrix} 0 \\ G \\ 0 \end{bmatrix} [g(x(t)) - g(\hat{x}(t))]
\tag{13.10}
$$

where

$$\bar{A} = \begin{bmatrix} W + LHE & 0 & 0 \\ HE & A - K & J \\ 0 & -\Gamma_2 & -\Gamma_1 \end{bmatrix}.$$

It is supposed that $\|F(t)\|^2 < \infty$, which is reasonable in most practical cases since usually $F(t)$ can only persist for a limited time, until the fault is diagnosed and the system is configured.

Definition 13.1 The reference output is defined as

$$z_\infty(t) = C_1 e_\omega(t) + C_2 e_x(t) + C_3 \tilde{F}(t) \tag{13.11}$$

Definition 13.2 For the constant $\gamma > 0$ and the weighting matrices $Q_1 > 0$ and $Q_2 > 0$, the generalized H_∞ performance is denoted as follows:

$$J_\infty = \|z_\infty\|^2 - \gamma^2 \|F\|^2 - \delta(Q_1, Q_2) \tag{13.12}$$

where

$$\delta(Q_1, Q_2) = e_x^T(0) Q_1 e_x(0) + e_\omega^T(0) Q_2 e_\omega(0)$$

At this stage, the objective is to find K, L and Γ_1, Γ_2 such that system (13.10) is stable.

Theorem 13.1

If for the parameter $\lambda_1 > 0$, matrices C_i ($i = 1, 2, 3$) and $Q_1 > 0, Q_2 > 0$, there exist matrices $P_1 > 0, P_2 > 0, R_1, R_2, \Gamma_1 > 0, \Gamma_2$ and constant $\gamma > 0$ satisfying $P_1 \leq Q_1, P_2 \leq Q_2$ and

$$\begin{bmatrix} \Xi_1 & E^T H^T P_1 & 0 & 0 & R_2 J & C_1^T \\ * & \Xi_2 & P_1 G & \Xi_{24} & 0 & C_2^T \\ * & * & -\frac{1}{\lambda_1^2} I & 0 & 0 & 0 \\ * & * & * & -2\Gamma_1 & \Gamma_1 & C_3^T \\ * & * & * & * & -\gamma^2 I & 0 \\ * & * & * & * & * & -I \end{bmatrix} < 0 \tag{13.13}$$

where

$$\Xi_1 = (P_2 W + R_2 HE) + (P_2 W + R_2 HE)^T,$$

$$\Xi_2 = (P_1 A - R_1) + (P_1 A - R_1)^T + \frac{1}{\lambda_1^2} U_1^T U_1$$

$$\Xi_{24} = P_1 J - \Gamma_2^T .$$

then with gains $K = P_1^{-1} R_1$ and $L = P_2^{-1} R_2$, error systems (13.10) are stable and satisfy $J_\infty < 0$. ∎

Proof. Let

$$V_1(t) = e_\omega^T(t) P_2 e_\omega(t) \tag{13.14}$$

$$
\begin{aligned}
V_2(t) &= e_x^T(t) P_1 e_x(t) + \tfrac{1}{\lambda_1^2} \int_0^t [\|U_1 e_x(\tau)\|^2 \\
&\quad - \|g(x(\tau)) - g(\hat{x}(\tau))\|^2] d\tau
\end{aligned}
\tag{13.15}
$$

$$V_3(t) = \widetilde{F}^T(t) \widetilde{F}(t) \tag{13.16}$$

Variable t will be omitted in the following procedures of proof to simplify, if there are no various interpretations.

Following (13.2) and (13.10), it is verified that $V_1(t) \geq 0$ and $V_2(t) \geq 0$ holds for all arguments. Along with the trajectories of (13.6), (13.8) and (13.9), it can be shown that

$$\dot{V}_1 = e_\omega^T [(P_2 W + P_2 LHE) + (P_2 W + P_2 LHE)^T] e_\omega + 2 e_\omega^T P_2 LJF$$

$$
\begin{aligned}
\dot{V}_2 &= e_x^T[(P_1 A - P_1 K) + (P_1 A - P_1 K)^T] e_x + 2 e_x^T P_1 G[g(x) - g(\hat{x})] \\
&\quad + \tfrac{1}{\lambda_1^2}[\|U_1 e_x\|^2 - \|g(x) - g(\hat{x})\|^2] + 2 e_x^T P_1 J \widetilde{F} + 2 e_x^T P_1 HE e_\omega
\end{aligned}
$$

$$\dot{V}_3 = -2 \widetilde{F}^T \Gamma_1 \widetilde{F} - 2 \widetilde{F}^T \Gamma_2 e_x + 2 \widetilde{F}^T \Gamma_1 F$$

A Lyapunov function candidate for (13.10) is chosen as

$$V(t) = V_1(t) + V_2(t) + V_3(t)$$

Consider an auxiliary function as the performance index

$$J_{aux} = \int_0^t (\|z_\infty(\tau)\|^2 - \gamma^2 \|F(\tau)\|^2 + \dot{V}) d\tau \tag{13.17}$$

It can be verified that

$$\|z_\infty\|^2 - \gamma^2 \|F\|^2 + \dot{V} = q^T \Psi_1 q \tag{13.18}$$

where

$$q^T = \begin{bmatrix} e_\omega^T & e_x^T & (g(x) - g(\hat{x}))^T & \widetilde{F}^T & F^T \end{bmatrix},$$

$$
\Psi_1 = \begin{bmatrix} \Xi_1 & E^T H^T P_1 & 0 & 0 & R_2 J \\ * & \Xi_2 & P_1 G & \Xi_{24} & 0 \\ * & * & -\frac{1}{\lambda_1^2} I & 0 & 0 \\ * & * & * & -2\Gamma_1 & \Gamma_1 \\ * & * & * & * & -\gamma^2 I \end{bmatrix}
$$

$$
+ \begin{bmatrix} C_1^T \\ C_2^T \\ 0 \\ C_3^T \\ 0 \end{bmatrix} [C_1 \ C_2 \ 0 \ C_3 \ 0] \tag{13.19}
$$

It can be seen by using the Schur complement formula that (13.13) leads to $\Psi_1 < 0$, and then $J_{aux} < 0$. It can be verified that

$$
J_\infty \leq J_{aux} + \delta(P_1 - Q_1, P_2 - Q_2)
$$

Since $\delta(P_1 - Q_1, P_2 - Q_2) < 0$, it can be obtained that $J_\infty < 0$. This completes the proof. $\qquad\square$

In the application of Theorem 13.1, it can be simply selected that $Q_i = P_i$, $(i = 1, 2)$ are the weighting matrices. Theorem 13.1 has actually provided a robust fault diagnosis design method.

In [90], the fault and its estimation were supposed to be less than a known value, which may be difficult to choose. In [96], the stability of fault estimation system was discussed. However, the stability of the fault estimation error system cannot be ensured. In this section, the generalized H_∞ optimization is applied to keep the stability of the fault estimation error.

13.4   SIMULATION EXAMPLES

In this section, we consider the longitudinal dynamics of an A4D aircraft at a flight condition of 15,000 ft altitude and Mach 0.9 given in [85]. The longitudinal dynamics can be denoted as

$$
\dot{x}(t) = Ax(t) + Gg(x(t)) + Hu(t) + JF(t) + Hd(t) \tag{13.20}
$$

where

$$
x(t) = [x_1(t) \ x_2(t) \ x_3(t) \ x_4(t)]^T
$$

$x_1(t)$ is the forward velocity (fts^{-1}), $x_2(t)$ is the angle of attack (rad), $x_3(t)$ is the pitching velocity $(rads^{-1})$, $x_4(t)$ is the pitching angle (rad), $u(t)$ is the elevator deflection (deg) and

$$
A_0 = \begin{bmatrix} -0.0605 & 32.37 & 0 & 32.2 \\ -0.00014 & -1.475 & 1 & 0 \\ -0.0111 & -34.72 & -2.793 & 0 \\ 0 & 0 & 1 & 0 \end{bmatrix}
$$

The state-feedback controller is $u(t) = K'x(t)$, where

$$K' = [2.3165\ 9.9455\ 4.0004\ 13.8525]$$

Then, we can conclude that

$$A = A_0 + HK'$$

$$= \begin{bmatrix} -0.0605 & 32.3700 & 0 & 32.2000 \\ -0.2466 & -2.5332 & 0.5744 & -1.4739 \\ -78.3088 & -370.8779 & -138.0065 & -468.2145 \\ 0 & 0 & 1.0000 & 0 \end{bmatrix}$$

$$H = \begin{bmatrix} 0 \\ -0.1064 \\ -33.8 \\ 0 \end{bmatrix}, J = \begin{bmatrix} 0 \\ 12 \\ 6 \\ 0 \end{bmatrix}.$$

It is supposed that

$$G = [0\ 0\ 50\ 0]^T, g(x(t)) = sin(2\pi 5t)x_2(t)$$

then, the matrix U_1 can be selected as $U_1 = diag\{0, 1, 0, 0\}$ and the norm condition (13.2) can be satisfied. $d(t)$ is assumed to be an unknown harmonic disturbance described by (13.3) with

$$W = \begin{bmatrix} 0 & 5 \\ -5 & 0 \end{bmatrix}, E = [25\ 0].$$

The initial values of the state and disturbance are supposed to be

$$x^T(0) = [0.6\ -0.6\ 0.9\ 0.6],$$

$$\hat{x}(0) = [0\ 0\ 0\ 0], \omega(0) = 0.1$$

For the reference output, it is denoted that

$$C_1 = [0.1\ 0.1], C_2 = [0.1\ 0.1\ 0.1\ 0.1], C_3 = 0.1$$

The fault is supposed to occur at the 15th second as $F(t) = 1$. For $\lambda_1^2 = 20$ and $\gamma = 1$, it can be solved via LMI related to (13.13) that

$$\Gamma_1 = 1.1396,$$

$$\Gamma_2 = [-0.7172\ 12.2516\ -0.0273\ -0.7172],$$

the gain of fault diagnosis observer is

$$K = 10^3 \times \begin{bmatrix} 0.0012 & 0.0184 & 0.0011 & 0.0166 \\ 0.0428 & -0.0134 & 0.0191 & -0.0214 \\ -1.0857 & -6.9258 & 0.6339 & -6.9394 \\ 0.0163 & 0.0025 & 0.0067 & 0.0004 \end{bmatrix}$$

and the gain of disturbance observer is

$$L = \begin{bmatrix} 0 & 0.0003 & 0.0016 & 0 \\ 0 & 0.0005 & 0.0003 & 0 \end{bmatrix}.$$

When the disturbance observer is constructed based on (13.4-13.5), the estimation error of exogenous disturbance is shown in Figure 13.1, where the tracking ability of the disturbance observer is satisfactory. The estimations of fault with exogenous disturbances are demonstrated in Figure 13.2, where the solid line represents the fault signal, the dash-dot line is its estimation in this chapter and the dotted line denotes its estimation based on generalized H_∞ optimization without disturbance rejection in [96]. Fault estimation error is shown in Figure 13.3. From Figures 13.2 and 13.3, it can be seen that the estimation of fault in this chapter is much more accurate than its counterpart in [96]. In Figure 13.4, the residual signal in the fault diagnosis observer is illustrated. In Figure 13.5, it is shown that the slowly time-varying fault $f = 1 + 0.1sin(1.2t)$ can also be well estimated, where the solid line represents the fault signal, the dash-dot line is its estimation in this chapter and dotted line is that in [96].

13.5   CONCLUSION

In this chapter, the fault diagnosis problem is investigated for nonlinear systems with external disturbances. The unknown external disturbances are supposed to be generated by an exogenous system. Disturbances are estimated by a disturbance observer, and can then be rejected in a fault diagnosis observer. A solution based on a linear matrix inequality is presented so that the fault estimation error system is stable and the fault can be estimated with disturbance rejection performance.

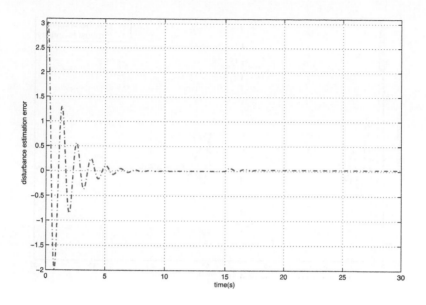

FIGURE 13.1   Disturbance estimation error in disturbance observer.

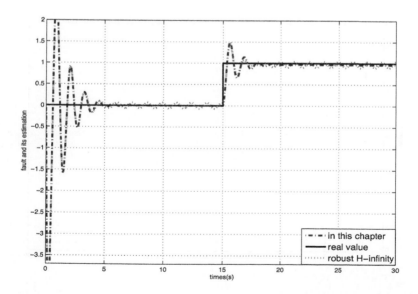

FIGURE 13.2   Constant fault and its estimation.

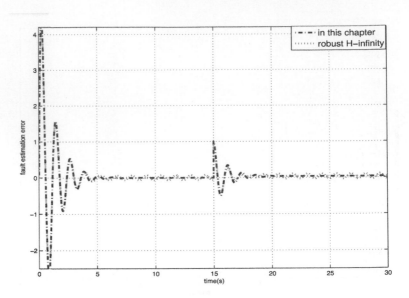

FIGURE 13.3   Fault estimation error in fault diagnosis observer.

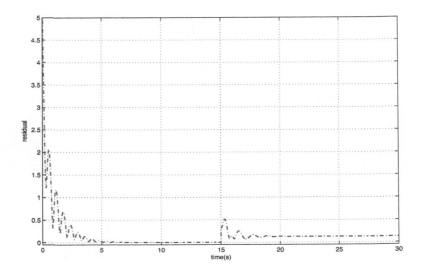

FIGURE 13.4   Residual signal in fault diagnosis observer.

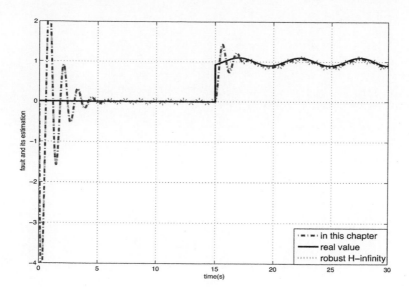

FIGURE 13.5   Slow-varying fault and its estimation.

14 Adaptive Fault Diagnosis Based on Disturbance Observer

14.1 INTRODUCTION

In [51], the generalized H_∞ optimization approach was applied to estimate the fault for stochastic distribution systems. In [236], a H_∞ controller in internal model architecture was presented to realize fault estimation and accommodation for linear systems with disturbances. The fault estimation and accommodation problem for multiple input multiple output (MIMO) stochastic systems was presented in [115], where the system noise was supposed to be Gaussian. It is worth pointing out that in most literature, the disturbances are assumed to be norm-bounded or Gaussian so that some robust control techniques (such as H_∞ or H_2) can be applied. In [71], an observer was proposed to simultaneously estimate system states and actuator faults, and to attenuate disturbances for descriptor systems. A composite observer was constructed in [19] by combining a fault diagnosis observer with a disturbance observer, with which a fault can be estimated and disturbances can be rejected at the same time. It is noted that a single disturbance was considered in the controlled systems in [19, 71], which would lead to conservatism of the proposed approaches.

In this chapter, the fault diagnosis problem is discussed for a class of systems with faults and multiple disturbances simultaneously. The disturbances are divided into two parts. The first part is formulated by an exogenous system and the second one is uncertain but bounded by an unknown nonnegative constant. It is shown that up to now, most previous results on fault diagnosis problems only considered faulty systems with no disturbances or only one disturbance [19, 71, 117, 236]. A composite observer for the modeled disturbance and time-varying faults is designed with an adaptive compensation term to reject unknown bounded disturbance. Additionally, the time-varying faults are considered in this chapter instead of the instant faults assumed in [19, 51].

14.2 PROBLEM FORMULATION AND PRELIMINARIES

Consider the following system with nonlinearities, faults and multiple disturbances

$$
\begin{cases}
\dot{x}(t) = & Ax(t) + Gg(x(t)) + H[u(t) + F(t)] \\
& + H_1 d_1(t) + H_2 d_2(t) \\
y(t) = & Cx(t)
\end{cases}
\tag{14.1}
$$

where $x(t) \in R^n$ is the system state, $u(t)$ is the control input, $y(t) \in R^m$ is output variable. $F(t)$ is the fault vector to be diagnosed. $F(t)$ is supposed to be time-varying but its derivative is assumed to be bounded. A, C, G, H, H_1 and H_2 represent the coefficient matrices of the system with suitable dimensions. The modeled external disturbance $d_1(t)$ is supposed to be generated by an exogenous system described by

$$\begin{cases} \dot{\omega}(t) &= W\omega(t) \\ d_1(t) &= E\omega(t) \end{cases} \tag{14.2}$$

where $\omega(t)$ is the state variable, $W \in R^{p \times p}$ and E are the known parameter matrices of the exogenous system. The uncertain disturbance $d_2(t)$ is supposed to be bounded with an unknown nonnegative constant a_d.

For a known matrix U_1, $g(x(t))$ is a known nonlinear vector function that is supposed to satisfy $g(0) = 0$ and the following norm condition

$$\|g(x_1(t)) - g(x_2(t))\| \le \|U_1(x_1(t) - x_2(t))\| \tag{14.3}$$

for any $x_1(t)$ and $x_2(t)$.

Remark 14.1 Multiple disturbances exist in many practical plants, such as unknown frictions or loads, harmonic disturbances, modeling uncertainties, environmental disturbance and measurement noises, etc. (see [20, 85]). Compared with the previous works [19, 85, 117, 236], both faults and multiple disturbances are considered in this chapter. It is worth pointing out that the constant a_d is an unknown nonnegative, which is supposed to be known in [19, 20, 22].

In the next section, the objective is to construct a new composite observer, with which the fault can be diagnosed, and the disturbances can be rejected and attenuated at the same time.

14.3   FAULT DIAGNOSIS OBSERVER DESIGN

14.3.1   DISTURBANCE OBSERVER

In this subsection, a disturbance observer will be presented to estimate the modeled external disturbance for a healthy system (i.e., $F(t) = 0$). By augmenting state Equation (14.1) with disturbance dynamics Equation (14.2), the composite system is given by

$$\begin{cases} \dot{X}(t) &= A_0 X(t) + G_0 g(X(t)) + H_0 u(t) \\ &\quad + H_{20} d_2(t) \\ y(t) &= C_0 X(t) \end{cases} \tag{14.4}$$

where

$$X(t) = \begin{bmatrix} x(t) \\ \omega(t) \end{bmatrix}, A_0 = \begin{bmatrix} A & H_1 E \\ 0 & W \end{bmatrix},$$

$$g(X(t)) = g(x(t)), C_0 = \begin{bmatrix} C & 0 \end{bmatrix},$$

$$H_0 = \begin{bmatrix} H \\ 0 \end{bmatrix}, G_0 = \begin{bmatrix} G \\ 0 \end{bmatrix}, H_{20} = \begin{bmatrix} H_2 \\ 0 \end{bmatrix}.$$

The following assumption is required so that the concerned problem can be well-posed in this chapter.

Assumption 14.1 (A_0, C_0) *is observable.*

The full-order observer for both $x(t)$ and $\omega(t)$ is designed as

$$\begin{cases} \dot{\hat{X}}(t) & = A_0\hat{X}(t) + G_0 g(\hat{X}(t)) + H_{20} u_a(t) \\ & \quad + H_0 u(t) + L(y(t) - \hat{y}(t)) \\ \hat{y}(t) & = C_0\hat{X}(t) \end{cases} \tag{14.5}$$

where $\hat{X}(t)$ is the estimation of state $X(t)$ and $\hat{y}(t)$ is the estimation of system output $y(t)$. Matrix L is the observer gain to be determined later and $u_a(t)$ is the compensation term formulated as

$$u_a(t) = \frac{J e_y(t)}{\|e_y^T(t) J^T\|} a_d \tag{14.6}$$

where $e_y(t) = y(t) - \hat{y}(t)$ is the estimation error of the output variable. J is a matrix with proper dimension to be determined later.

Defining

$$e_X(t) = X(t) - \hat{X}(t), \tilde{g}(t) = g(X(t)) - g(\hat{X}(t)),$$

the estimation error system can be obtained from (14.4) and (14.5) to show

$$\begin{aligned} \dot{e}_X(t) & = (A_0 - LC_0) e_X(t) + G_0 \tilde{g}(t) \\ & \quad + H_{20}(d_2(t) - u_a(t)) \end{aligned} \tag{14.7}$$

At this stage, the objective is to find observer gain L such that system (14.7) is stable.

Theorem 14.1

If for the parameter $\lambda > 0$, matrices U, there exist matrices $P > 0$, R satisfying

$$\begin{bmatrix} sym(PA_0 - RC_0) + \lambda U^T U & PG_0 \\ * & -\lambda I \end{bmatrix} < 0 \tag{14.8}$$

and

$$JC_0 = H_{20}^T P \tag{14.9}$$

then with gains $L = P^{-1}R$, the estimation error system (14.7) is asymptotically stable.
∎

Proof. The Lyapunov function candidate for system (14.7) is chosen as

$$
\begin{aligned}
V(t) \;&=\; e_x^T(t)Pe_x(t) \\
&\quad + \lambda \int_0^t [\|Ue_x(\tau)\|^2 - \|\tilde{g}(\tau)\|^2]d\tau
\end{aligned}
\tag{14.10}
$$

Following (14.7), it is verified that $V(t) \geq 0$ holds for all arguments. Along with the trajectories of (14.7), its derivative can be shown that

$$
\begin{aligned}
\dot{V}(t) \;&=\; e_X^T(t)sym(PA_0 - PLC_0)e_X(t) \\
&\quad + 2e_X^T(t)PG_0\tilde{g}(t) + \lambda[\|Ue_x(t)\|^2 - \|\tilde{g}(t)\|^2] \\
&\quad + 2e_X^T(t)PH_{20}(d_2(t) - u_a(t)) \\
&=\; e_X^T(t)[sym(PA_0 - PLC_0) + \lambda U^T U]e_X(t) \\
&\quad + 2e_X^T(t)PG_0\tilde{g}(t) - \lambda\|\tilde{g}(t)\|^2 \\
&\quad + 2e_x^T(t)P_1H_{20}(d_2(t) - u_a(t)) \\
&\leq\; e_X^T(t)[sym(PA_0 - PLC_0) + \lambda U^T U]e_X(t) \\
&\quad + 2e_X^T(t)PG_0\tilde{g}(t) - \lambda\|\tilde{g}(t)\|^2 + 2\|e_X^T(t)PH_{20}\|a_d \\
&\quad - 2e_X^T(t)PH_{20}\frac{Je_y(t)}{\|e_y^T(t)J^T\|}a_d \\
&=\; e_X^T(t)[sym(PA_0 - PLC_0) + \lambda U^T U]e_X(t) + 2e_X^T(t)PG_0\tilde{g}(t) \\
&<\; 0
\end{aligned}
$$

It can be seen that the estimation error system (14.7) is asymptotically stable. This completes the proof. □

It is noted that parameter a_d needs to be determined in the observer, which requires certain information on the external disturbance. The next step is to design adaptive estimation for the unknown constant a_d. Then, the compensation term (14.6) can be further formulated as follows

$$u_a(t) = \frac{Je_y(t)}{\|e_y^T(t)J^T\|}\hat{a}_d(t) \tag{14.11}$$

with

$$\dot{\hat{a}}_d(t) = \frac{1}{\gamma} \|e_y^T(t)J^T\| \tag{14.12}$$

where $\hat{a}_d(t)$ is the estimation of the constant a_d, γ is the selected adaptive rate. The design scheme for the full-order observer can be further formulated as follows.

Theorem 14.2

If for the parameter $\lambda > 0$, matrix U, there exist matrices $P > 0$, R satisfying (14.8) and (14.9), then with gains $L = P^{-1}R$, the estimation error system (14.7) is asymptotically stable with the adaptive compensation terms (14.11) and (14.12). ∎

Proof. At the end of the proof Theorem 14.1, it can be seen that

$$
\begin{aligned}
\dot{V}(t) &= e_X^T(t)[sym(PA_0 - PLC_0) + \lambda U^T U]e_X(t) \\
&\quad + 2e_X^T(t)PG_0\tilde{g}(t) - \lambda \|\tilde{g}(t)\|^2 \\
&\quad + 2e_x^T(t)P_1H_{20}(d_2(t) - u_a(t)) \\
&\leq e_X^T(t)[sym(PA_0 - PLC_0) + \lambda U^T U]e_X(t) \\
&\quad + 2e_X^T(t)PG_0\tilde{g}(t) - \lambda \|\tilde{g}(t)\|^2 \\
&\quad + 2\|e_X^T(t)PH_{20}\|(a_d - \hat{a}_d)
\end{aligned}
$$

Selecting the Lyapunov function as

$$V_0(t) = V(t) + \gamma(a_d - \hat{a}_d)^2 \tag{14.13}$$

Then, it can be seen that

$$
\begin{aligned}
\dot{V}_0(t) &= \dot{V}(t) - 2\gamma(a_d - \hat{a}_d)\dot{\hat{a}}_d \\
&\leq e_X^T(t)[sym(PA_0 - PLC_0) + \lambda U^T U]e_X(t) \\
&\quad + 2e_X^T(t)PG_0\tilde{g}(t) - \lambda \|\tilde{g}(t)\|^2 \\
&\quad + 2\|e_X^T(t)PH_{20}\|(a_d - \hat{a}_d) - 2\gamma(a_d - \hat{a}_d)\dot{\hat{a}}_d \\
&\leq e_X^T(t)[sym(PA_0 - PLC_0) + \lambda U^T U]e_X(t) \\
&\quad + 2e_X^T(t)PG_0\tilde{g}(t) - \lambda \|\tilde{g}(t)\|^2 \\
&\quad + 2\|e_X^T(t)PH_{20}\|(a_d - \hat{a}_d) \\
&\quad - 2\gamma(a_d - \hat{a}_d)\frac{1}{\gamma}\|e_y^T(t)J^T\|
\end{aligned}
$$

$$= e_X^T(t)[sym(PA_0 - PLC_0) + \lambda U^T U]e_X(t)$$
$$+2e_X^T(t)PG_0\tilde{g}(t) - \lambda \|\tilde{g}(t)\|^2$$
$$< 0$$

It can be concluded that the estimation error (14.7) is asymptotically stable with the adaptive compensation terms (14.11) and (14.12). This completes the proof. □

Remark 14.2 In this chapter, the bounded disturbance $d_2(t)$ is compensated for by an adaptive scheme, which was attenuated by a generalized H_∞ optimization approach in [19, 20]. Moreover, the estimation error system is asymptotically stable for the proposed method in this chapter.

14.3.2   FAULT DIAGNOSIS OBSERVER

Fault estimation needs to be accomplished in order to diagnosis the fault. For this purpose, the following composite observer will be constructed to estimate the system state, fault and modeled uncertain disturbance simultaneously. By augmenting the system state with disturbance and fault dynamics, the composite system can be further given by

$$\begin{cases} \dot{\bar{X}}(t) &= \bar{A}\bar{X}(t) + \bar{G}\tilde{g}(t) + \bar{H}u(t) + \bar{I}\dot{F}(t) + \bar{H}_2 d_2(t) \\ y(t) &= \bar{C}\bar{X}(t) \end{cases} \tag{14.14}$$

where

$$\bar{X}(t) = \begin{bmatrix} x(t) \\ \omega(t) \\ F(t) \end{bmatrix}, \bar{A} = \begin{bmatrix} A & H_1E & H \\ 0 & W & 0 \\ 0 & 0 & 0 \end{bmatrix},$$

$$\bar{I} = \begin{bmatrix} 0 \\ 0 \\ I \end{bmatrix}, \bar{H}_2 = \begin{bmatrix} H_2 \\ 0 \\ 0 \end{bmatrix}, \bar{G} = \begin{bmatrix} G \\ 0 \\ 0 \end{bmatrix},$$

$$\bar{H} = \begin{bmatrix} H \\ 0 \\ 0 \end{bmatrix}, \bar{C} = \begin{bmatrix} C & 0 & 0 \end{bmatrix}.$$

Assumption 14.2 $(\bar{A}, \bar{C})$ *is observable.*

For the system (14.14), the composite observer is formulated as

$$\begin{cases} \dot{\hat{\bar{X}}}(t) &= \bar{A}\hat{\bar{X}}(t) + \bar{G}\hat{\tilde{g}}(t) + \bar{H}u(t) \\ &\quad + \bar{L}(y(t) - \hat{y}(t)) + \bar{H}_2 u_a(t) \\ \hat{y}(t) &= \bar{C}\hat{\bar{X}}(t) \end{cases} \tag{14.15}$$

where $\hat{\bar{X}}^T(t) = [\begin{array}{ccc} \hat{x}^T(t) & \hat{\omega}^T(t) & \hat{F}^T(t) \end{array}]$ is the estimation of $\bar{X}(t)$, $\hat{\bar{g}}(t)$ represents the estimation of $\bar{g}(t)$. $u_a(t)$ is the adaptive compensation term defined in (14.11). $\bar{L}^T = [\begin{array}{ccc} L_1^T & L_2^T & L_3^T \end{array}]$ is the observer gain to be determined later.

Denoting

$$\bar{e}(t) = \bar{X}(t) - \hat{\bar{X}}(t), \tilde{\bar{g}}(t) = \bar{g}(t) - \hat{\bar{g}}(t),$$

the estimation error system yields

$$\dot{\bar{e}}(t) \quad = \quad (\bar{A} - \bar{L}\bar{C})\bar{e}(t) + \bar{G}\tilde{\bar{g}}(t) + \bar{I}\dot{F}(t)$$

$$+ \bar{H}_2(d_2(t) - u_a(t)) \tag{14.16}$$

Remark 14.3 In [19, 51], the faults are supposed to be constants, which is a restrictive condition. Time-varying faults are considered in this chapter. The fault change ratio is assumed to be bounded, which is reasonable in most practical cases since usually $F(t)$ can only persist for a limited time, until the fault is diagnosed and the system is configured.

At this stage, the objective is to find $\bar{L}$ such that system (14.16) is stable. The following result provides a new design method for the fault diagnosis problem based on LMIs.

Theorem 14.3

If for the parameter $\bar{\lambda} > 0$, matrix $\bar{U}$, there exist matrices $\bar{P} > 0$, $\bar{R}$ and constant $\eta > 0$, satisfying

$$\begin{bmatrix} sym(\bar{P}\bar{A} - \bar{R}\bar{C}) + \bar{\lambda}\bar{U}^T\bar{U} + \eta I & \bar{P}\bar{G} \\ * & -\bar{\lambda}I \end{bmatrix} < 0 \tag{14.17}$$

and

$$J\bar{C} = \bar{H}_2^T\bar{P} \tag{14.18}$$

then with gains $\bar{L} = \bar{P}^{-1}\bar{R}$, error system (14.16) is stable, and the error satisfies

$$\|\bar{e}(t)\| \leq max\{\|\bar{e}(0)\|, 2\eta^{-1}a_F\|\bar{P}\bar{I}\|\} \tag{14.19}$$

for all $t \in [0, +\infty)$.   ∎

Proof. Let

$$\bar{V}(t) = \quad \bar{e}^T(t)\bar{P}\bar{e}(t) + \gamma(a_d - \hat{a}_d(t))^2$$

$$+ \bar{\lambda}\int_0^t[\|\bar{U}\bar{e}(\tau)\|^2 - \|\tilde{\bar{g}}(\tau)\|^2]d\tau \tag{14.20}$$

Following the Equation (14.16), it is verified that $\bar{V}(t) \geq 0$ holds for all arguments. Along with the trajectories of (14.16), it can be shown that

$$
\begin{aligned}
\dot{\bar{V}}(t) \;=\; & \bar{e}^T(t)sym(\bar{P}\bar{A}-\bar{P}\bar{L}\bar{C})\bar{e}(t)+2\bar{e}^T(t)\bar{P}\bar{G}\tilde{\bar{g}}(t) \\
& +2\bar{e}^T(t)\bar{P}\bar{I}\dot{F}(t)+\bar{\lambda}[\|\bar{U}\bar{e}(t)\|^2-\|\tilde{\bar{g}}(t)\|^2] \\
& +2\bar{e}^T(t)\bar{P}\bar{H}_2 d_2(t)-2\bar{e}^T(t)\bar{P}\bar{H}_2 u_a(t) \\
& -2\gamma(a_d-\hat{a}_d(t))\frac{1}{\gamma}\|e_y^T(t)J^T\| \\[4pt]
\leq\; & \bar{e}^T(t)[sym(\bar{P}\bar{A}-\bar{P}\bar{L}\bar{C})+\bar{\lambda}\bar{U}^T\bar{U}]\bar{e}(t) \\
& +2\bar{e}^T(t)\bar{P}\bar{G}\tilde{\bar{g}}(t)+2\|\bar{e}^T(t)\bar{P}\bar{H}_2\|(a_d-\hat{a}_d) \\
& -\bar{\lambda}\|\tilde{\bar{g}}(t)\|^2-2\gamma(a_d-\hat{a}_d)\frac{1}{\gamma}\|e_y^T(t)J^T\| \\
& +2\bar{e}^T(t)\bar{P}\bar{I}\dot{F}(t) \\[4pt]
\leq\; & \bar{e}^T(t)[sym(\bar{P}\bar{A}-\bar{P}\bar{L}\bar{C})+\bar{\lambda}\bar{U}^T\bar{U}]\bar{e}(t) \\
& +\frac{1}{\bar{\lambda}}\bar{e}^T(t)\bar{P}\bar{G}\bar{G}^T\bar{P}\bar{e}(t)+2\bar{e}^T(t)\bar{P}\bar{I}\dot{F}(t)
\end{aligned}
$$

By denoting $\bar{R}=\bar{P}\bar{L}$ and using Schur complement formula with respect to (14.17), we have $\Psi < -\eta I$, where

$$
\Psi = sym(\bar{P}\bar{A}-\bar{P}\bar{L}\bar{C})+\bar{\lambda}\bar{U}^T\bar{U}+\frac{1}{\bar{\lambda}}\bar{P}\bar{G}\bar{G}^T\bar{P} \tag{14.21}
$$

It can be concluded that

$$
\begin{aligned}
\dot{\bar{V}}(t) \;\leq\; & -\eta\|\bar{e}(t)\|^2+2\bar{e}^T(t)\bar{P}\bar{I}\dot{F}(t) \\
\leq\; & -\eta\|\bar{e}(t)\|^2+2\|\bar{e}(t)\|\ \|\bar{P}\bar{I}\|\ \|\dot{F}(t)\| \\
\leq\; & -\eta\|\bar{e}(t)\|^2+2a_F\|\bar{e}(t)\|\ \|\bar{P}\bar{I}\|
\end{aligned}
$$

which means that $\bar{V}(t)<0$ if $\|\bar{e}(t)\|>2a_F\|\bar{P}\bar{I}\|$ occurs. Therefore, it can be verified that (14.19) always holds and the estimation error system is stable. This completes the proof. □

If the fault is assumed to be constant (i.e., $\dot{F}(t)=0$), the result of asymptotical stability for (14.16) can be obtained as given in the following corollary.

Corollary 14.1 If for the parameter $\bar{\lambda}>0$ and matrix $\bar{U}$, there exist matrices $\bar{P}>0,\bar{R}$ and constant $\eta>0$ satisfying (14.17) and (14.18), then with gain $\bar{L}=\bar{P}^{-1}\bar{R}$, the error system (14.16) is asymptotically stable when the fault is assumed to be constant.

Proof. At the end of proof of Theorem 14.3, when $\dot{F}(t) = 0$, it is shown that

$$\dot{V}(t) \leq \bar{e}^T(t)[sym(\bar{P}\bar{A} - \bar{P}\bar{L}\bar{C}) + \bar{\lambda}\bar{U}^T\bar{U}]\bar{e}(t) + \frac{1}{\lambda}\bar{e}^T(t)\bar{P}\bar{G}\bar{G}^T\bar{P}\bar{e}(t)$$

$$\leq -\eta \parallel \bar{e}(t) \parallel^2$$

This implies that the error system is asymptotically stable. This completes the proof. □

14.4 SIMULATION EXAMPLES

In this section, we consider the longitudinal dynamics of an A4D aircraft at a flight condition of 15,000 ft altitude and Mach 0.9 given in [19, 85]. The longitudinal dynamics can be denoted as

$$\begin{cases} \dot{x}(t) &= A_1x(t) + Gg(x(t)) + H_1u(t) \\ &\quad + J_2F(t) + H_1d_1(t) + H_2d_2(t) \\ y(t) &= Cx(t) \end{cases} \quad (14.22)$$

where the state variable is

$$x(t) = \begin{bmatrix} x_1(t) & x_2(t) & x_3(t) & x_4(t) \end{bmatrix}^T$$

$x_1(t)$ is the forward velocity (fts^{-1}), $x_2(t)$ is the angle of attack (rad), $x_3(t)$ is the pitching velocity $(rads^{-1})$, $x_4(t)$ is the pitching angle (rad), $u(t)$ is the elevator deflection (deg). The coefficient matrices of aircraft model are given by

$$A_1 = \begin{bmatrix} -0.0605 & 32.37 & 0 & 32.2 \\ -0.00014 & -1.475 & 1 & 0 \\ -0.0111 & -34.72 & -2.793 & 0 \\ 0 & 0 & 1 & 0 \end{bmatrix}$$

$$C = I_4, H_1 = \begin{bmatrix} 0 \\ -0.1064 \\ -33.8 \\ 0 \end{bmatrix}, H_2 = \begin{bmatrix} 0.1 \\ 0 \\ -3 \\ 0.1 \end{bmatrix}.$$

The state-feedback controller is $u(t) = K'x(t)$, where

$$K' = \begin{bmatrix} 2.3165 & 9.9455 & 4.0004 & 13.8525 \end{bmatrix}$$

Then, we can conclude that

$$A = A_1 + HK'$$

$$= \begin{bmatrix} -0.0605 & 32.3700 & 0 & 32.2000 \\ -0.2466 & -2.5332 & 0.5744 & -1.4739 \\ -78.3088 & -370.8779 & -138.0065 & -468.2145 \\ 0 & 0 & 1.0000 & 0 \end{bmatrix}$$

Since there may be a large degree of nonlinearity and/or uncertainty in the (3, 2) entry of A (see [85]), it is supposed that

$$G = [\ 0\ \ 0\ \ 50\ \ 0\ ]^T,$$

$$g(x(t)) = sin(2\pi 5t)x_2(t)$$

then, the matrix U_1 can be selected as $U_1 = diag\{0,\ 1,\ 0,\ 0\}$ and the norm condition (14.3) can be satisfied. Periodic disturbance $d_1(t)$ caused by rotating aerial propeller is assumed to be an unknown harmonic disturbance described by (14.2) with

$$W = \begin{bmatrix} 0 & 5 \\ -5 & 0 \end{bmatrix},\ E = [\ 25\ \ 0\ ].$$

Wind gust and system noises $d_2(t)$ can be considered as a random signal with bounded upper 2-norm.

The initial values of the states are supposed to be

$$x^T(0) = [\ 2\ \ -2\ \ 3\ \ 2\ ]$$

For $\lambda = 10$, it can be solved via LMI related to (14.17) that the gain of composite observer (15) is

$$J = [\ 0.4177\ \ \ 0.0000\ \ \ -0.1239\ \ \ 0.4177\ ],$$

and

$$\bar{L} = \begin{bmatrix} 0.9178 & 14.4268 & -1.2129 & 16.5400 \\ 17.7481 & 0.6239 & -2.9905 & -1.2170 \\ 44.3902 & -9.6741 & -33.8085 & -93.6539 \\ 15.6600 & -0.2635 & -2.7015 & 0.9783 \\ -0.3005 & -3.4941 & -8.5975 & -0.8514 \\ -0.0573 & -0.6531 & -1.5827 & -0.1577 \\ 1.4245 & 12.1549 & -0.2603 & -0.0828 \end{bmatrix},$$

When the disturbance observer is constructed, the estimation error of exogenous disturbances is shown in Figure 14.1. The actuator bias fault is supposed to occur at the 15th second as $F = 4(deg)$. The estimations of fault with system disturbances are demonstrated in Figure 14.2, where the solid line represents the real fault signal, the dash line is its estimation in this chapter and the dash-dot line denotes its estimation based on generalized H_∞ with disturbance observer in [19]. Fault estimation error is shown in Figure 14.3. From Figures 14.2 and 14.3, the estimation of fault in this chapter is much more accurate than its counterpart in [19]. In Figure 14.4, it is shown that the time-varying actuator fault $F = 4 + 0.5sin(2t)(deg)$ can also be well estimated, where the solid line represents the fault signal, the dash line is its estimation in this chapter and dash-dot line is generalized H_∞ with disturbance observer in [19]. The estimation error is shown in Figure 14.5. From Figures 14.4 and 14.5, the adaptive fault diagnosis approach is more accurate than its counterpart in [19] for time-varying faults, which can show the efficiency of the proposed approach.

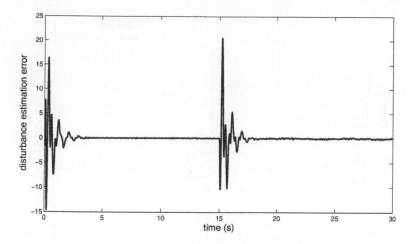

FIGURE 14.1   Disturbance estimation error in disturbance observer.

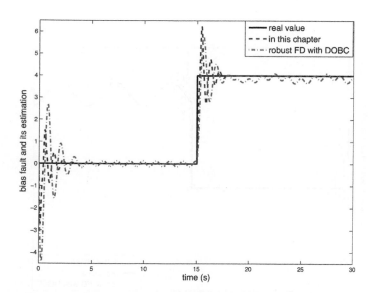

FIGURE 14.2   Actuator bias fault and its estimations.

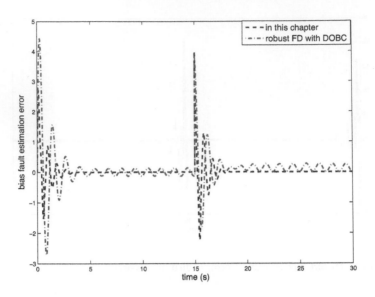

FIGURE 14.3 Fault estimation errors in fault diagnosis observer.

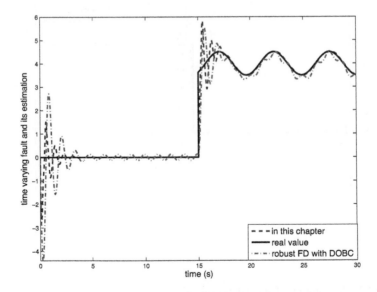

FIGURE 14.4 Time-varying actuator fault and its estimations.

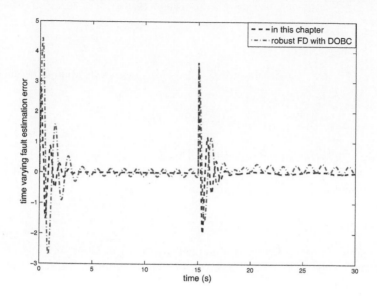

FIGURE 14.5   Estimation of time varying fault.

14.5   CONCLUSION

In this chapter, the anti-disturbance fault diagnosis problem is investigated for systems with faults and multiple disturbances. The multiple disturbances are supposed to include two different variables. The first one is described by an exogenous model and the second one is uncertain but bounded by an unknown constant. Modeled exogenous disturbances and time-varying faults are estimated by a composite observer with adaptive compensation. Finally, we apply this approach into an A4D aircraft flight control system. It is shown that with the proposed anti-disturbance observer, fault diagnosis performance can be improved.

15 Anti-Disturbance FTC for Systems with Multiple Disturbances

15.1 INTRODUCTION

Fault-tolerant design approaches can be broadly classified into two types: passive fault-tolerant control (FTC) and active FTC. The passive approaches are relatively conservative and the desired control performance for the closed-loop systems may not be satisfied, especially when the number of possible faults and the degrees of system redundancy increase [119, 224]. An active FTC approach can reconfigure systems either by selecting a pre-computed control law or by synthesizing an online control strategy; among these the observer design and robust control methodologies have been widely applied (see [71, 115, 116, 232]).

The FTC problem will be much more complicated due to the existence of faults and disturbances simultaneously. To overcome these obstacles, some elegant approaches have been provided and applied in practical engineering (see [69, 231, 236]). In [231], an optimal FTC scheme was studied for nonlinear systems. Generalized H_∞ optimization was applied to estimate the fault and to attenuate the disturbances resulting from the modeling processes and stochastic noises. In [70], an observer was proposed to simultaneously estimate system states and actuator faults to attenuate disturbances for descriptor systems. In [236], a H_∞ controller in the internal model architecture was presented to realize fault estimation and accommodation for linear systems with disturbances. It is noted that in most literature, the disturbances are assumed to be norm-bounded so that some robust control techniques such as H_∞ optimization can be applied.

In this chapter, a composite fault-tolerant controller is addressed for faulty systems with multiple disturbances based on disturbance observer based control (DOBC) methods. The disturbances are divided into two parts, which are modeled as exogenous disturbances and norm-bounded uncertain disturbances. A composite controller is designed combining a fault diagnosis observer and disturbance observer, with which the fault can be diagnosed with enhanced disturbance rejection and attenuation performance. It is shown that up to now, most previous results on fault tolerance problems only considered faulty systems with no disturbances or only one disturbance [231, 236]. In this chapter multiple disturbances are studied for fault tolerant control problems, where the situation will turn to be more complex. Additionally, the time-varying faults are considered in this chapter instead of the instant faults assumed in [19, 231].

15.2 PROBLEM FORMULATION AND PRELIMINARIES

Consider the following nonlinear system with faults and multiple disturbances

$$\begin{cases} \dot{x}(t) &= Ax(t) + Gg(x(t)) + H_1[u(t) + F(t)] \\ &\quad + H_1 d_1(t) + H_2 d_2(t) \\ y(t) &= Cx(t) \end{cases} \tag{15.1}$$

where $x(t) \in R^m$ is the system state, $u(t)$ is the control input, $y(t)$ is output variable. $F(t)$ is the actuator fault vector to be diagnosed. $F(t)$ is supposed to be time-varying but its derivative is supposed to be bounded. A, C, G, H_1 and H_2 represent the coefficient matrices of the weight system with suitable dimensions.

The modeled external disturbance $d_1(t)$ is supposed to be generated by a linear exogenous system described by

$$\begin{cases} \dot{\omega}(t) &= W\omega(t) + H_3\delta(t) \\ d_1(t) &= E\omega(t) \end{cases} \tag{15.2}$$

where $\omega(t)$ is the state variable, $W \in R^{p \times p}$, E and H_3 are the known parameter matrices of the exogenous system. $\delta(t)$ is the additional disturbance which results from the perturbations and uncertainties in the exogenous system. The disturbances $d_2(t)$ and $\delta(t)$ are supposed to have the bounded H_2 norm.

For a known matrix U_1, $g(x(t))$ is a known nonlinear vector function that is supposed to satisfy $g(0) = 0$ and the following norm condition

$$\|g(x_1(t)) - g(x_2(t))\| \leq \|U_1(x_1(t) - x_2(t))\| \tag{15.3}$$

for any $x_1(t)$ and $x_2(t)$.

The following assumptions are required so that the problem concerned can be wellposed in this chapter.

Assumption 15.1 (A, H_1) *is controllable,* (W, H_1E) *is observable.*

Remark 15.1 Multiple disturbances exist in many practical plants, such as unknown frictions or loads, harmonic disturbances, modeling uncertainties and stochastic noises, etc. (see [85, 201]). In this chapter, the multiple disturbances are supposed to include two parts, which are uncertain modeled disturbances (such as periodic disturbances in vibrating structures, rotating mechanisms with eccentricity) and norm-bounded uncertain disturbances (such as modeling uncertainties, perturbations and stochastic noises). Compared with previous works [85, 231, 236], both faults and multiple disturbances are considered simultaneously.

In the next section, the objective is to construct a new composite robust controller, with which the fault can be accommodated, and the disturbances can be rejected and attenuated at the same time. The design idea can be depicted in Figure 15.1.

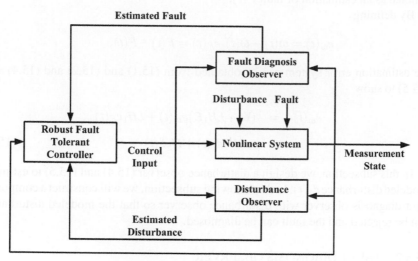

FIGURE 15.1   The block diagram of anti-disturbance FTC system.

15.3   ROBUST FAULT TOLERANT CONTROLLER DESIGN

15.3.1   DISTURBANCE OBSERVER

In order to reject the modeled external disturbance, a disturbance observer is designed in this subsection. In [85], a full-order observer is designed to estimate disturbances and states when the states are unavailable. The reduced order observer is constructed to estimate disturbances when the states are available. In this note, we only consider the case of available states. When all states of a system are available, it is unnecessary to estimate the states, and only the estimation of the disturbance need be considered. The disturbance observer is formulated as

$$\begin{cases} \hat{\omega}(t) &= r(t) - Lx(t) \\ \hat{d}_1(t) &= E\hat{\omega}(t) \end{cases} \tag{15.4}$$

where $r(t)$ is the auxiliary variable generated by

$$\dot{r}(t) = (W + LH_1E)[r(t) - Lx(t)] + L[Ax(t) \\ + Gg(x(t)) + H_1u(t) + H_1u_{fc}(t)] \tag{15.5}$$

$\hat{\omega}(t)$ is the estimation of $\omega(t)$, $\hat{d}_1(t)$ is the estimation of modeled disturbance $d_1(t)$, matrix L is the disturbance observer gain to be determined later and $u_{fc}(t) = \hat{F}(t)$ is the compensation term to be designed in the fault diagnosis observer, where $\hat{F}(t)$ is denoted as an estimation of fault $F(t)$.

By defining

$$e_\omega(t) = \omega(t) - \hat{\omega}(t), e_F(t) = F(t) - \hat{F}(t),$$

the estimation error system can be obtained from (15.1) and (15.2) and (15.4) and (15.5) to show

$$\begin{aligned} \dot{e}_\omega(t) &= (W + LH_1E)e_\omega(t) + LH_1e_F(t) \\ &\quad + LH_2d_2(t) + H_3\delta(t) \end{aligned} \tag{15.6}$$

In this subsection, we design a disturbance observer (15.4) and (15.5) to estimate modeled disturbance $d_1(t)$. In the following subsection, we will construct a composite fault diagnosis observer with disturbance observer so that the modeled disturbance can be rejected and the fault can be diagnosed.

15.3.2 FAULT DIAGNOSIS OBSERVER

Fault estimation needs to be accomplished in order to reconfigure a system. For this purpose, the following fault diagnosis observer is constructed to diagnose the fault.

$$\begin{cases} \hat{F}(t) &= \xi(t) - Kx(t) \\ \dot{\xi}(t) &= KH_1(\xi(t) - Kx(t)) + K[Ax(t) + Gg(x(t)) \\ &\quad + H_1(u(t) + u_{dc}(t))] \end{cases} \tag{15.7}$$

where $\hat{F}(t)$ is the estimation of $F(t)$. The DOBC term $u_{dc}(t) = \hat{d}_1(t)$ is applied to reject modeled disturbance $d_1(t)$ by its estimation from disturbance observer. K is the fault diagnosis observer gain to be determined later.

The fault estimation error system yields

$$\begin{aligned} \dot{e}_F(t) &= \dot{F}(t) - \dot{\hat{F}}(t) \\ &= \dot{F}(t) + KH_1e_F(t) + KH_1Ee_\omega(t) + KH_2d_2(t) \end{aligned} \tag{15.8}$$

Remark 15.2 In [19, 231], the faults are supposed to be constants, which is a restrictive condition. Time-varying faults are considered in this chapter. The fault change ratio is assumed to be bounded, which is reasonable in most practical cases since usually $F(t)$ can only persist for a limited time, until the fault is diagnosed and the system is configured.

In this subsection, a fault diagnosis observer is designed for fault estimation. In the next subsection, a composite fault-tolerant controller will be determined for reconfiguring systems with disturbance rejection and attenuation performance.

15.3.3 COMPOSITE FAULT TOLERANT CONTROLLER

The intention is to design a control scheme to guarantee that the system (15.1) is stable in the presence (or absence) of faults and multiple disturbances simultaneously for the nonlinear system concerned in this section. The structure of the composite fault-tolerant controller is formulated as

$$u(t) = u_{sc}(t) - u_{fc}(t) - u_{dc}(t) \tag{15.9}$$

where

$$u_{fc}(t) = \hat{F}(t), u_{dc}(t) = \hat{d}_1(t), u_{sc}(t) = Mx(t)$$

M is the state feedback controller gain to be determined later. Substituting (15.9) into (15.1), it can be seen that

$$
\begin{aligned}
\dot{x}(t) &= (A + H_1 M)x(t) + Gg(x(t)) + H_1 e_F(t) \\
&\quad + H_1 E e_\omega(t) + H_2 d_2(t)
\end{aligned}
\tag{15.10}
$$

Variable t will be omitted in the following procedures for simplification, if there are no various interpretations.

Combing estimation error Equation (15.6) with Equations (15.8) and (15.10) yields

$$
\begin{bmatrix} \dot{x} \\ \dot{e}_\omega \\ \dot{e}_F \end{bmatrix}
= \bar{A} \begin{bmatrix} x \\ e_\omega \\ e_F \end{bmatrix}
+ \begin{bmatrix} Gg(x) \\ 0 \\ 0 \end{bmatrix}
+ \begin{bmatrix} 0 \\ H_3 \\ 0 \end{bmatrix} \delta
$$
$$
+ \begin{bmatrix} 0 \\ 0 \\ I \end{bmatrix} \dot{F}
+ \begin{bmatrix} H_2 \\ LH_2 \\ KH_2 \end{bmatrix} d_2
\tag{15.11}
$$

where

$$
\bar{A} = \begin{bmatrix}
A + H_1 M & H_1 E & H_1 \\
0 & W + LH_1 E & LH_1 \\
0 & KH_1 E & KH_1
\end{bmatrix}.
$$

After the modeled disturbance and fault are rejected by their estimations, H_∞ optimization technique will be applied to attenuate the norm-bounded uncertain disturbances.

Definition 15.1 The reference output is defined as

$$z_\infty(t) = C_0 x(t) + C_1 e_\omega(t) + C_2 e_F(t) \tag{15.12}$$

where $C_i, (i = 0, 1, 2)$ are the selected weighting matrices.

Definition 15.2 For the constant $\gamma_1 > 0$, $\gamma_2 > 0$, $\gamma_3 > 0$ and the weighting matrices $Q_0 > 0$, and $Q_2 > 0$, the generalized H_∞ performance is denoted as follows:

$$J_\infty = \|z_\infty\|^2 - \gamma_1^2\|\dot{F}\|^2 - \gamma_2^2\|d_2\|^2 - \gamma_3^2\|\delta\|^2 - \rho(Q_0, Q_2) \tag{15.13}$$

where

$$\rho(Q_0, Q_2) = x^T(0)Q_0^{-1}x(0) + \begin{bmatrix} e_\omega^T(0) & 0 \end{bmatrix} Q_2 \begin{bmatrix} e_\omega^T(0) & 0 \end{bmatrix}^T$$

At this stage, the objective is to find K, L and M such that system (15.11) is stable. The following result provides a new design method for the FTC problem based on LMIs (linear matrix inequalities).

Theorem 15.1

If for the parameter $\lambda > 0$, matrices C_i ($i = 0, 1, 2$) and $Q_0 > 0$, $Q_2 > 0$, there exist matrices $P_0 > 0$, $P_2 > 0$, R_0, R_2 and constants $\gamma_1 > 0$, $\gamma_2 > 0$, $\gamma_3 > 0$ satisfying $P_0^{-1} \leq Q_0^{-1}$, $P_2 \leq Q_2$ and

$$\begin{bmatrix} \Xi_1 & H_1\bar{E} & G & \Xi_{14} & \Xi_{15} & \Xi_{16} \\ * & \Xi_2 & 0 & \Xi_{24} & 0 & \bar{C}^T \\ * & * & -\frac{1}{\lambda^2}I & 0 & 0 & 0 \\ * & * & * & \Xi_4 & 0 & 0 \\ * & * & * & * & -\lambda^2 I & 0 \\ * & * & * & * & * & -I \end{bmatrix} < 0 \tag{15.14}$$

where

$$\Xi_1 = sym(AP_0 + H_1R_0), \quad \Xi_{14} = \begin{bmatrix} 0 & H_2 & 0 \end{bmatrix},$$

$$\Xi_{15} = P_0U_1^T, \quad \Xi_{16} = P_0C_0^T,$$

$$\Xi_2 = sym(P_2\bar{W} + R_2H_1\bar{E}),$$

$$\Xi_{24} = \begin{bmatrix} P_2\bar{H}_1 & R_2H_2 & P_2\bar{H}_3 \end{bmatrix},$$

$$\Xi_4 = \begin{bmatrix} -\gamma_1^2 I & 0 & 0 \\ 0 & -\gamma_2^2 I & 0 \\ 0 & 0 & -\gamma_3^2 I \end{bmatrix},$$

$$\bar{W} = \begin{bmatrix} W & 0 \\ 0 & 0 \end{bmatrix}, \bar{E} = \begin{bmatrix} E & I \end{bmatrix},$$

$$\bar{H}_1 = \begin{bmatrix} 0 \\ I \end{bmatrix}, \ \bar{H}_3 = \begin{bmatrix} H_3 \\ 0 \end{bmatrix}, \ \bar{C} = \begin{bmatrix} C_1 & C_2 \end{bmatrix}.$$

then with gains $M = R_0 P_0^{-1}$ and $\begin{bmatrix} L \\ K \end{bmatrix} = P_2^{-1} R_2$, error system (15.11) is stable and satisfy $J_\infty < 0$. ∎

Proof. Let

$$V_1(t) = x^T(t)Px(t) + \frac{1}{\lambda^2} \int_0^t [\|U_1 x(\tau)\|^2 - \|g(x(\tau))\|^2]d\tau \qquad (15.15)$$

$$V_2(t) = \bar{e}^T(t) P_2 \bar{e}(t) \qquad (15.16)$$

where $\bar{e}(t) = \begin{bmatrix} e_\omega^T(t) & e_F^T(t) \end{bmatrix}^T$. Following (15.11), it is verified that $V_1 \geq 0$ and $V_2 \geq 0$ holds for all arguments. Along with the trajectories of (15.11), it can be shown that

$$\begin{aligned}
\dot{V}_1 &= x^T sym(PA + PH_1 M)x + 2x^T PGg(x) + \frac{1}{\lambda_0^2}[\|U_1 x\|^2 - \|g(x)\|^2] \\
&\quad + 2x^T PH_1 \bar{E}\bar{e} + 2x^T PH_2 d_2
\end{aligned}$$

$$\begin{aligned}
\dot{V}_2 &= \bar{e}^T sym(P_2 \bar{W} + P_2 \begin{bmatrix} L \\ K \end{bmatrix} H_1 \bar{E})\bar{e} + 2\bar{e}^T P_2 \begin{bmatrix} L \\ K \end{bmatrix} H_2 d_2 \\
&\quad + 2\bar{e}^T P_2 \bar{H}_3 \delta + 2\bar{e}^T P_2 \bar{H}_1 \dot{F}
\end{aligned}$$

A Lyapunov function candidate for system (15.11) is chosen as

$$V(t) = V_1(t) + V_2(t)$$

Consider an auxiliary function as the performance index

$$\begin{aligned}
J_{aux} &= \int_0^t (\|z_\infty(\tau)\|^2 - \gamma_1^2 \|\dot{F}(\tau)\|^2 - \gamma_2^2 \|d_2(\tau)\|^2 \\
&\quad - \gamma_3^2 \|\delta(\tau)\|^2 + \dot{V}(\tau))d\tau
\end{aligned} \qquad (15.17)$$

It can be verified that

$$\|z_\infty\|^2 - \gamma_1^2 \|\dot{F}\|^2 - \gamma_2^2 \|d_2\|^2 - \gamma_3^2 \|\delta\|^2 + \dot{V} = q^T \Psi_1 q \qquad (15.18)$$

where

$$q^T = \begin{bmatrix} x^T & \bar{e}^T & g^T(x) & \dot{F}^T & d_2^T & \delta^T \end{bmatrix},$$

$$\Psi_1 = \begin{bmatrix} \Psi_{11} & PH_1\bar{E} & PG & \Psi_{14} \\ * & \Psi_{22} & 0 & \Psi_{24} \\ * & * & -\frac{1}{\lambda^2}I & 0 \\ * & * & * & \Xi_4 \end{bmatrix}$$

$$+ \begin{bmatrix} C_0^T \\ \bar{C}^T \\ 0 \\ 0 \end{bmatrix} \begin{bmatrix} C_0 & \bar{C} & 0 & 0 \end{bmatrix} \tag{15.19}$$

where

$$\Psi_{11} = sym(PA + PH_1M) + \frac{1}{\lambda^2}U_1^T U_1,$$

$$\Psi_{14} = \begin{bmatrix} 0 & PH_2 & 0 \end{bmatrix},$$

$$\Psi_{22} = sym(P_2\bar{W} + P_2\begin{bmatrix} L \\ K \end{bmatrix} H_1\bar{E}),$$

$$\Psi_{24} = \begin{bmatrix} P_2\bar{H}_1 & R_2H_2 & P_2\bar{H}_3 \end{bmatrix},$$

It can be seen by using the Schur complement formula that (15.14) leads to $\Psi < 0$, where

$$\Psi = diag\{P^{-1},I,I,I\}^T \Psi_1 diag\{P^{-1},I,I,I\}$$

and $P^{-1} = P_0$. It can be further concluded that $\Psi_1 < 0$, and then $J_{aux} < 0$. It can be verified that

$$J_\infty \leq J_{aux} + \rho(Q_0 - P_0, P_2 - Q_2)$$

Since $\rho(Q_0 - P_0, P_2 - Q_2) < 0$, it can be obtained that $J_\infty < 0$. This completes the proof.   □

In the application of Theorem 15.1, it can be simply selected that $P_i = Q_i, (i = 0, 2)$ are the weighting matrices. Theorem 15.1 has thus provided a robust fault-tolerant controller design method.

If the norm-bounded uncertain disturbances $d_2(t)$ and $\delta(t)$ are neglected (i.e., $d_2(t) = 0$ and $\delta(t) = 0$), the result of asymptotical stability for (15.11) can be obtained as given in the following corollary, when the fault is assumed to be constant.

Corollary 15.1 If for the parameter $\lambda > 0$, there exist matrices $P_0 > 0$, $P_2 > 0$, R_0 and R_2 satisfying

$$\Phi = \begin{bmatrix} \Xi_1 & H_1\bar{E} & G & P_0U_1^T \\ * & \Xi_2 & 0 & 0 \\ * & * & -\frac{1}{\lambda^2}I & 0 \\ * & * & * & -\lambda^2 I \end{bmatrix} < 0 \tag{15.20}$$

then with gains $M = R_0 P_0^{-1}$, $\begin{bmatrix} L \\ K \end{bmatrix} = P_2^{-1} R_2$, error system (15.11) is asymptotically stable when $d_2(t) = 0$, $\delta(t) = 0$ and $\dot{F}(t) = 0$.

Proof. At the end of proof of Theorem 15.1, when $d_2(t) = 0$, $\delta(t) = 0$ and $\dot{F}(t) = 0$, it is shown that

$$\dot{V}_1 = x^T[(PA + PH_1 M) + (PA + PH_1 M)^T]x + 2x^T PGg(x)$$
$$+ \frac{1}{\lambda_0^2}[\|U_1 x\|^2 - \|g(x)\|^2] + 2x^T PH_1 \bar{E}\bar{e}$$

$$\dot{V}_2 = \bar{e}^T sym(P_2 \bar{W} + P_2 \begin{bmatrix} L \\ K \end{bmatrix} H_1 \bar{E})\bar{e}$$

Furthermore, it can be seen that

$$\dot{V}(t) = q_1^T(t) \Phi_1 q_1(t)$$

where

$$q_1^T(t) = \begin{bmatrix} x^T & \bar{e}^T(t) & g^T(x(t)) \end{bmatrix}$$

It can be seen by using the Schur complement formula that $\Phi < 0$ leads to $\Phi_1 < 0$. This implies that the error system is asymptotically stable. This completes the proof. $\square$

15.4 SIMULATION EXAMPLES

In this section, we consider the longitudinal dynamics of an A4D aircraft at a flight condition of 15,000 ft altitude and Mach 0.9 given in [85]. The longitudinal dynamics can be denoted as

$$\dot{x}(t) = Ax(t) + Gg(x(t)) + H_1(u(t) + F(t)) + H_1 d_1(t) + H_2 d_2(t) \tag{15.21}$$

where $x(t) = \begin{bmatrix} x_1(t) & x_2(t) & x_3(t) & x_4(t) \end{bmatrix}^T$ are measurable by using sensors technique.. $x_1(t)$ is the forward velocity (fts^{-1}), $x_2(t)$ is the angle of attack (rad), $x_3(t)$ is the pitching velocity $(rads^{-1})$, $x_4(t)$ is the pitching angle (rad), $u(t)$ is the elevator deflection (deg). The coefficient matrices of the aircraft model are given by

$$A = \begin{bmatrix} -0.0605 & 32.37 & 0 & 32.2 \\ -0.00014 & -1.475 & 1 & 0 \\ -0.0111 & -34.72 & -2.793 & 0 \\ 0 & 0 & 1 & 0 \end{bmatrix}$$

$$H_1 = \begin{bmatrix} 0 \\ -0.1064 \\ -33.8 \\ 0 \end{bmatrix}, H_2 = \begin{bmatrix} 0.1 \\ 0 \\ -3 \\ 0.1 \end{bmatrix}.$$

Since there may be a large degree of nonlinearity and/or uncertainty in the $(3, 2)$ entry of A (see [85]), it is supposed that

$$G = [\ 0 \quad 0 \quad 50 \quad 0\ ]^T, \ g(x(t)) = sin(2\pi 5t)x_2(t)$$

then, the matrix U_1 can be selected as $U_1 = diag\{0,\ 1,\ 0,\ 0\}$ and the norm condition (15.3) can be satisfied. Periodic disturbance $d_1(t)$ caused by rotating aerial propeller is assumed to be an unknown harmonic disturbance described by (15.2) with

$$W = \begin{bmatrix} 0 & 5 \\ -5 & 0 \end{bmatrix}, \ E = [\ 25 \quad 0\ ], \ H_3 = \begin{bmatrix} 0.1 \\ 0.1 \end{bmatrix}.$$

$\delta(t)$ is the additional disturbance signal resulting from the perturbations and uncertainties in the exogenous system (15.2), and satisfies 2-norm boundedness. In simulation, we select $\delta(t)$ as the random signal with upper 2-norm bound 1. Wind gust and system noises $d_2(t)$ can also be considered as a random signal with bounded upper 2-norm.

The initial values of the states are supposed to be

$$x^T(0) = [\ 2 \quad -2 \quad 3 \quad 2\ ]$$

For the reference output, it is denoted that

$$C_0 = [\ 1 \quad 1 \quad 1 \quad 1\ ], C_1 = [\ 1 \quad 1\ ], C_2 = 1.$$

For $\lambda = 1, \gamma_1 = 1, \gamma_2 = 1$ and $\gamma_3 = 1$, it can be solved via LMI related to (15.14) that the gain of fault diagnosis observer (15.7) is

$$K = [\ 14.4090 \quad -1.8344 \quad 0.9571 \quad 14.4788\ ],$$

the gain of disturbance observer (15.4) is

$$L = \begin{bmatrix} 0.1785 & 0.1226 & 0.0120 & 0.1762 \\ 1.1450 & -0.0072 & 0.0762 & 1.1475 \end{bmatrix},$$

and the gain of state feedback controller is

$$M = [\ 14.1731 \quad 14.6596 \quad 14.0713 \quad 15.7827\ ].$$

When the disturbance observer is constructed based on (15.4) and (15.5), the estimation error of exogenous disturbances is shown in Figure 15.2. The actuator bias fault is supposed to occur at the 15th second as $F = 4$. The estimations of fault with system disturbances are demonstrated in Figure 15.3, where the solid line represents the real fault signal, the dashed line is its estimation in this chapter and the dash-dot line denotes its estimation based on generalized H_∞ optimization without disturbance rejection in [231]. Fault estimation error is shown in Figure 15.4. From Figures 15.3 and 15.4, the estimation of fault in this chapter is much more accurate than its counterpart in [231]. In Figure 15.5, the state response signals of the control system are illustrated. It can be seen that the proposed fault-tolerant controller has a good control ability for both configuring faults, and rejecting and attenuating disturbances simultaneously. In Figure 15.6, the ramp fault is assumed to occur at the 15th second with slope 0.1. In Figure 15.7, it is shown that the time-varying fault $F = 4 + 0.5 sin(2t)$ can also be well estimated, where the solid line represents the fault signal, the dashed line is its estimation in this chapter and dash-dot line generalized H_∞ in [231]. From Figures 15.6 and 15.7, the estimations of fault with disturbance rejection are much more accurate than their counterpart of robust H_∞ in [231] for time-varying faults, which can show the efficiency of the proposed approach.

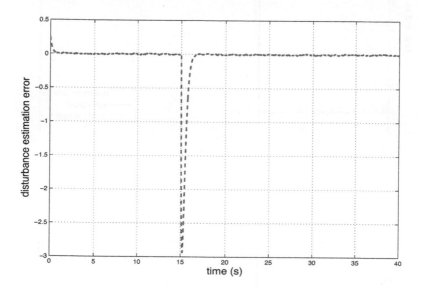

FIGURE 15.2   Disturbance estimation error in disturbance observer.

15.5   CONCLUSION

In this chapter, the anti-disturbance FTC problem is investigated for nonlinear systems with faults and multiple disturbances. The multiple disturbances are supposed to include uncertain modeled exogenous disturbances and norm-bounded uncertain disturbances. Modeled exogenous disturbances are estimated by a disturbance observer, and fault is diagnosed in a fault diagnosis observer. A composite controller is designed integrating fault accommodation from a diagnosis observer with disturbance rejection from a disturbance observer, with which the fault can be accommodated and the disturbances can be rejected and attenuated simultaneously.

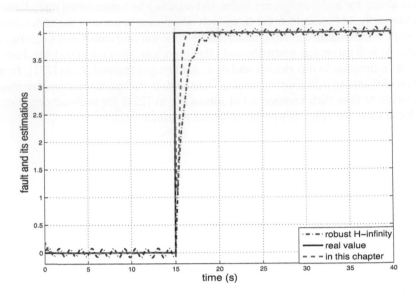

FIGURE 15.3   Bias fault and its estimations.

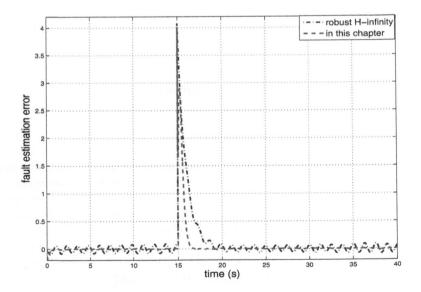

FIGURE 15.4   Fault estimation errors in fault diagnosis observer.

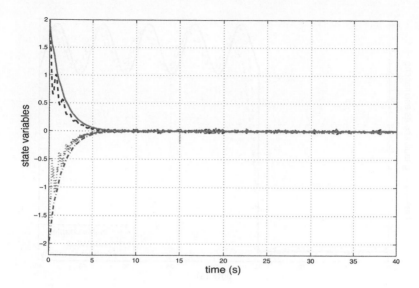

FIGURE 15.5   The responses of state variable.

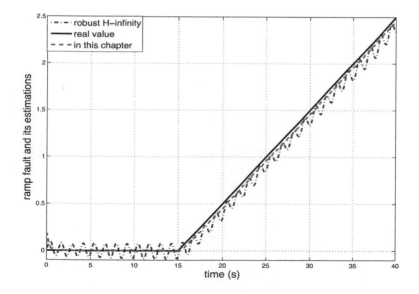

FIGURE 15.6   Ramp fault and its estimations.

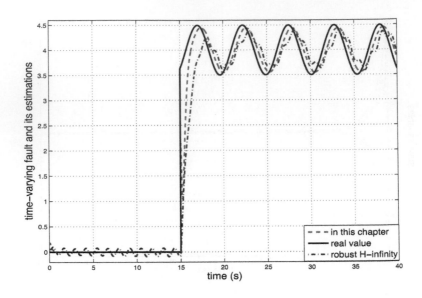

FIGURE 15.7   Time-varying fault and its estimations.

References

1. H. S. Ahn and C. H. Won. Fast alignment using rotation vector and adaptive kalman filter. *IEEE Trans. Aero. Elec. Sys.*, 42(1):70–83, 2006.
2. K. J. Astrom. *Introduction to Stochastic Control Theory*. Academic Press, New York, 1970.
3. K. J. Astrom and B. Wittenmark. *Adaptive Control*. Addison-Wesley, Boston, MA, 2nd edition, 1994.
4. J. Back and H. Shim. Adding robustness to nominal output feedback controllers for uncertain nonlinear systems: a nonlinear version of disturbance observer. *Automatica*, 44(9):2528–2537, 2008.
5. S. L. Ballois and G. Duc. H_∞ control of an earth observation satellite. *J. Guid. Control Dyn.*, 19(3):628–635, 1996.
6. I. Y. Bar-Itzhack and N. Berman. Control theoretic approach to inertial navigation systems. *J. Guid. Control Dyn.*, 11(3):237–245, 1988.
7. T. Basar and P. Bernhard. H_∞ *Optimal Control and Related Minimax Design Problems*. Birkhauser, Boston, 1995.
8. M. Basseville and I. Nikiforov. Fault isolation for diagnosis: nuisance rejection and multiple hypotheses testing. *Annu. Rev. Control*, 26:189–202, 2002.
9. J. Ben-Asher, J. A. Burns, and E. M. Cliff. Time optimal slewing of flexible spacecraft. *J. Guid. Control Dyn.*, 15:360–367, 1992.
10. R. Bickel and R. Tomizuka. Passivity-based versus disturbance observer based robot control: equivalence and stability. *J. Dyn. Syst-T ASME*, 121:41–47, 1999.
11. P. A. Bliman and G. Ferrari-Trecate. Average consensus problems in networks of agents with delayed communications. *Automatica*, 44(8):1985–1995, 2008.
12. M. Bodson and S. C. Douglas. Adaptive algorithms for the rejection of sinusoidal disturbances with unknown frequency. *Automatica*, 33(12):2213–2221, 1997.
13. M. Bodson, J. S. Jensen, and S. C. Douglas. Active noise control for periodic disturbances. *IEEE Trans. Contr. Syst. Technol.*, 9(1):200–205, 2001.
14. A. A. Bolonkin and N. S. Khot. Optimal bounded control design for vibration suppression. *Acta Astronaut.*, 38(10):803–813, 1996.
15. B. Brogliato and A. T. Neto. Practical stabilization of a class of nonlinear systems with partially known uncertainties. *Automatica*, 31:145–150, 1995.
16. L. J. Brown and Q. Zhang. Periodic disturbance cancellation with uncertain frequency. *Automatica*, 40(4):631–637, 2004.
17. C. I. Byrnes, F. D. Priscoli, and A. Isidori. *Output Regulation of Uncertain Nonlinear Systems*. Springer, Boston, 1997.
18. K. W. Byun, B. Wie, and D. Geller. Robust H_∞ control design for the space station with structured parameter uncertainty. *J. Guid. Control Dyn.*, 14(6):1115–1122, 1996.
19. S. Y. Cao and L. Guo. Fault diagnosis with disturbance rejection performance based on disturbance observer. pages 6947–6951, Shanghai, China, 2009. Joint

48th IEEE CDC & 28th CCC.

20. S. Y. Cao and L. Guo. Robust fault diagnosis with disturbance rejection and attenuation for systems with multiple disturbances. *J. Syst. Eng. Electron. Technol.*, 22(1):135–140, 2011.

21. S. Y. Cao and L. Guo. Multi-objective robust initial alignment algorithm for inertial navigation system with multiple disturbances. *Aerosp. Sci. Technol.*, 21(1):1–6, 2012.

22. S. Y. Cao, L. Guo, and X. Y. Wen. Fault tolerant control with disturbance rejection and attenuation performance for systems with multiple disturbances. *Asian J. Control*, 13(6):1056–1064, 2011.

23. Y. Cao and J. Lam. Robust H_∞ control of uncertain markovian jump systems with time delay. *IEEE Trans. Automat. Contr.*, 45(1):77–83, 2000.

24. S. P. Chan. A disturbance observer for robot manipulators with application to electronic components assembly. *IEEE Trans. Ind. Electron.*, 42:487–493, 1995.

25. C. Charbonnel. H_∞ and LMI attitude control design: towards performances and robustness enhancement. *Acta Astronaut.*, 54:307–314, 2004.

26. J. Chen and R. J. Patton. Optimal filtering and robust fault diagnosis of stochastic systems with unknown disturbances. *IEE P. Contr. Theor. Ap.*, 143:31–36, 1996.

27. M. Chen and W. H. Chen. Sliding mode control for a class of uncertain nonlinear system based on disturbance observer. *Int. J. Adapt. Control*, 24:51–64, 2010.

28. W. H. Chen. Nonlinear PID predictive control of two-link robotic manipulators. pages 217–222, Vienna, Austria, 2000. Proc. of the IFAC Symposium on Robot Control.

29. W. H. Chen. Nonlinear disturbance observer based control for nonlinear systems with harmonic disturbances. pages 316–321. In Proc. IFAC Symp NCSD' 01, 2001.

30. W. H. Chen. Nonlinear disturbance observer-enhanced dynamic inversion control of missiles. *J. Guid. Control Dynam.*, 26(1):161–166, 2003.

31. W. H. Chen. A nonlinear harmonic observer for nonlinear systems. *J. Dyn. Syst-T ASME*, 125:114–134, 2003.

32. W. H. Chen. Disturbance observer based control for nonlinear systems. *IEEE/ASME Trans. Mech.*, 9(4):706–710, 2004.

33. W. H. Chen and L. Guo. Analysis of disturbance observer based control for nonlinear systems under disturbances with bounded variation. pages ID–048, University of Bath, UK, 2004. In Proc. the Control Conf.

34. W. T. Chen and M. Saif. Actuator fault diagnosis for affine nonlinear systems with unknown inputs. pages 360–365, New Orleans, USA, 2007. In Proc. of the 46th IEEE CDC.

35. X. Chen, S. Komada, and T. Fukuda. Design of a nonlinear disturbance observer. *IEEE Trans. Ind. Electron.*, 47:429–437, 2000.

36. X. K. Chen, C. Y. Su, and F. Toshio. A nonlinear disturbance observer for multivariable systems and its application to magnetic bearing systems. *IEEE Trans. Contr. Syst. Technol.*, 12(4):569–577, 2004.

37. Y. P. Chen and S. C. Lo. Sliding mode controller design for spacecraft attitude tracking maneuvers. *IEEE Trans. Aero. Elec. Sys.*, 29(4):1328–1333, 1993.
38. M. Chilali and P. Gahinet. H_∞ design with pole placement constraints: An LMI approach. *IEEE Trans. Autom. Contr.*, 41(3):358–367, 1996.
39. B. K. Choi, W. K. Choi, and H. Lim. Nonlinear predictive PID controller. *IEEE/ASME Trans. Mech.*, 4:157–168, 1999.
40. J. H. Chou, I. R. Hong, and B. S. Chen. Dynamical feedback compensator for uncertain time-delay systems containing saturating actuator. *Int. J. Control*, 49:961–968, 1989.
41. D. Y. Chung, C. G. Park, and J. G. Lee. Observability analysis of strapdown inertial navigation system using lyapunov transformation. pages 23–28, WA02, Kobe, 1996. in Proc. of the 35th CDC.
42. J. L. Crassidis and F. L. Markley. Sliding mode control using modified rodrigues parameters. *J. Guid. Control Dyn.*, 19(6):1381–1383, 1996.
43. E. J. Davison. The robust control of a servo mechanism problem for linear time-invariable multivariable systems. *IEEE Trans. Automat. Contr.*, 21:25–34, 1976.
44. A. J. Van der Schaft. L_2-gain analysis of nonlinear systems and nonlinear state feedback H_∞ control. *IEEE Trans. Automat. Contr.*, 37:770–784, 1992.
45. Z. Ding. Universal disturbance rejection for nonlinear systems in output feedback form. *IEEE Trans. Automat. Contr.*, 48(7):1222–1226, 2003.
46. Z. T. Ding. Global stabilization and disturbance suppression of a class of nonlinear systems with uncertain internal model. *Automatica*, 39(3):471–479, 2003.
47. Z. T. Ding. Asymptotic rejection of asymmetric periodic disturbances in output-feedback nonlinear systems. *Automatica*, 43:555–561, 2007.
48. G. Y. Cheng et al. A microdrive track following controller design using robust and perfect tracking control with nonlinear compensation. *Mechatronics*, 15:933–948, 2005.
49. K. H. Kim et al. A current control for a permanent magnet synchronous motor with a simple disturbance estimation scheme. *IEEE Trans. Contr. Syst. tech.*, 7(5):630–633, 1999.
50. K. M. Lynch et al. Decentralized environmental modeling by mobile sensor networks. *IEEE Trans. Robot.*, 24(3):710–724, 2008.
51. L. Guo et al. Observer-based optimal fault detection and diagnosis using conditional probability distributions. *IEEE Trans. Signal Process.*, 54:3712–3719, 2006.
52. M. Y. Zhong et al. An LMI approach to design robust fault detection filter for uncertain LTI systems. *Automatica*, 39:543–550, 2003.
53. R. M. Murray et al. Future directions in control in an information-rich world. *IEEE Contr. Syst. Mag.*, 23:20–33, 2003.
54. S. H. Yu et al. Continuous finite-time control for robotic manipulators with terminal sliding mode. *Automatica*, 41:1957–1964, 2005.
55. S. Y. Xu et al. A delay-dependent approach to robust H_∞ filtering for uncertain distributed delay systems. *IEEE Trans. Signal Process.*, 53(10):3764–3772,

2005.

56. T. Floquet et al. On the robust fault detection via a sliding mode disturbance observer. *Int. J. Control*, 77:622–629, 2004.

57. W. H. Chen et al. Non-linear predictive PID controller. *IEE P. Contr. Theor. Ap.*, 146:603–611, 1999.

58. W. H. Chen et al. A nonlinear disturbance observer for robotic manipulators. *IEEE Trans. Ind. Electron.*, 47(4):932–938, 2000.

59. Z. D. Wang et al. Robust H_∞ filtering for stochastic time-delay systems with missing measurements. *IEEE Trans. Signal Process.*, 54(7):2579–2587, 2006.

60. J. C. Fang and D. J. Wan. A fast initial alignment method for strapdown inertial navigation system stationary base. *IEEE Trans. Aero. Elec. Sys.*, 32(4):1501–1505, 1996.

61. J. A. Farrell and M. Barth. *The Global Positioning System & Inertial Navigation*. McGraw -Hall companies Inc., USA, 1999.

62. J. A. Fax and R. M. Murray. Information flow and cooperative control of vehicle formations. *IEEE Trans. Automat. Contr.*, 49(9):1465–1475, 2004.

63. C. B. Feng and K. J. Zhang. *Robust Control for Nonlinear Systems (in Chinese)*. Science Press, Beijing, 2004.

64. T. Floquet and J. P. Barbot. State and unknown input estimation for linear discrete-time systems. *Automatica*, 42:1883–1889, 2006.

65. B. Friedl and Y. J. Park. On adaptive friction compensation. *IEEE Trans. Automat. Contr.*, 37:1609–1612, 1992.

66. H. J. Gao and T. W. Chen. New results on stability of discrete-time systems with time-varying state delay. *IEEE Trans. Automat. Contr.*, 52(2):328–334, 2007.

67. H. J. Gao and C. Wang. A delay-dependent approach to robust H_∞ filtering for uncertain discrete-time state-delayed systems. *IEEE Trans. Signal Process.*, 52:1631–1640, 2004.

68. Z. Q. Gao, Y. Huang, and J. Q. Han. An alternative paradigm for control system design. pages 4578–4585, Orlando, USA, 2001. 40th IEEE CDC.

69. Z. W. Gao and S. X. Ding. Actuator fault robust estimation and fault-tolerant control for a class of nonlinear descriptor systems. *Automatica*, 43:912–920, 2007.

70. Z. W. Gao and S. X. Ding. Sensor fault reconstruction and sensor compensation for a class of nonlinear state-space systems via descriptor system approach. *IET Control Theory Appl.*, 1(3):578–585, 2007.

71. Z. W. Gao and S. X. Ding. State and disturbance estimator for time-delay systems with application to fault estimation and signal compensation. *IEEE Trans. Signal Process.*, 55(12):5541–5551, 2007.

72. Z. W. Gao and D. W. C. Ho. State/noise estimator for descriptor systems with application to sensor fault diagnosis. *IEEE Trans. Signal Process.*, 54(4):1316–1326, 2006.

73. Z. W. Gao and H. Wang. Descriptor observer approaches for multivariable systems with measurement noises and application in fault detection and diagnosis. *Syst. Control Lett.*, 55:304–313, 2006.

74. S. D. Gennaro. Active vibration suppression in flexible spacecraft attitude tracking. *J. Guid. Control Dyn.*, 21(3):400–408, 1998.

75. S. D. Gennaro. Output stabilization of flexible spacecraft with active vibration suppression. *IEEE Trans. Aero. Elec. Sys.*, 39(3):747–759, 2003.

76. J. J. Gertler. *Fault Detection and Diagnosis in Engineering Systems*. Marcel Dekker, New York, 1998.

77. M. S. Grewal, L. R. Weill, and A. P. Andrews. *Global Positioning Systems, Inertial Navigation, and Integration*. John Wiley & Sons, Inc., New Jersey, USA, 2nd edition, 2007.

78. P. D. Groves. Optimising the transfer alignment of weapon INS. *J. Navigation*, 56:323–335, 2003.

79. G. O. Guardabassi and S. M. Savaresi. Approximate linearization via feedback an overview. *Automatica*, 37:1–15, 2001.

80. L. Guo. Guaranteed cost control of uncertain discrete-time delay systems using dynamic output feedback. *Trans. Inst. Meas. Control*, 24(5):417–430, 2002.

81. L. Guo. H_∞ output feedback control for delay systems with nonlinear and parametric uncertainties. *IEE P. Contr. Theor. Ap.*, 149:226–236, 2002.

82. L. Guo and S. Y. Cao. Initial alignment for nonlinear inertial navigation systems with multiple disturbances based on enhanced anti-disturbance filtering. *Int. J. Control*, 85(5):491–501, 2012.

83. L. Guo and W. H. Chen. DOB control for dynamic nonlinear systems with disturbances: a survey and comparisons. pages 105–110, Nottingham, U.K., 2001. Proc of CACSCUK01.

84. L. Guo and W. H. Chen. Output feedback H_∞ control for a class of uncertain nonlinear discrete-time delay systems. *Trans. Inst. Meas. Control*, 25(2):107–121, 2003.

85. L. Guo and W. H. Chen. Disturbance attenuation and rejection for systems with nonlinearity via DOBC approach. *Int. J. Robust Nonlin.*, 15(3):109–125, 2005.

86. L. Guo and D. Z. Cheng. *An Introduction to Control Theory - From Basic Concepts to the Research Frontier (in Chinese)*. Science Press, Beijing, 2005.

87. L. Guo, C. B. Feng, and W. H. Chen. A survey of disturbance observer-based control for dynamic nonlinear system. *Dyn. Contin. Discrete and Impulsive Syst-B*, 13E:79–84, 2006.

88. L. Guo and M. Malabre. Robust H_∞ control for descriptor systems with nonlinear uncertainties. *Int. J. Control*, 76(12):1254–1262, 2003.

89. L. Guo and M. Tomizuka. High-speed and high-precision motion control with an optimal hybrid feedforward controller. *IEEE/ASME Trans. Mech.*, 2:110–122, 1997.

90. L. Guo and H. Wang. Fault detection and diagnosis for general stochastic systems using b-spline expansions and nonlinear filters. *IEEE Trans. Circuits-I*, 52(8):1644–1652, 2005.

91. L. Guo and H. Wang. *Stochastic Distribution Control System Design: A Convex Optimization Approach*. Springer, London, 2009.

92. L. Guo, H. Wang, and T. Chai. Fault detection for non-linear non-gaussian

stochastic systems using entropy optimization principle. *Trans. Inst. Measm. Control*, 28:145–161, 2006.

93. L. Guo and X. Y. Wen. Hierarchical anti-disturbance adaptive control for nonlinear systems with composite disturbances and applications to missile systems. *Trans. Inst. Meas. Control*, 33(8):942–956, 2011.

94. L. Guo, X. Y. Wen, and X. Xin. Hierarchical composite anti-disturbance control for robotic systems using robust disturbance observer. pages 229–243. Robot Intelligence, Advanced Information and Knowledge Proc., 2010.

95. L. Guo, F. W. Yang, and J. C. Fang. Multiobjective filtering for nonlinear time-delay systems with nonzero initial conditions based on convex optimization. *Circ. Syst. Signal Process.*, 25(5):591–607, 2006.

96. L. Guo and Y. M. Zhang. Generalized robust H_∞ fault diagnosis filtering based on conditional stochastic distributions of system outputs. *J. Syst. Control Engineering*, 221:857–864, 2007.

97. X. Y. Guo and M. Bodson. Analysis and implementation of an adaptive algorithm for the rejection of multiple sinusoidal disturbances. *IEEE Trans. Contr. Syst. Technol.*, 17(1):40–50, 2009.

98. J. Q. Han. From PID to active disturbance rejection control. *IEEE Trans. Ind. Electron.*, 56(3):900–906, 2009.

99. S. L. Han and J. L. Wang. A novel initial alignment scheme for low-cost INS aided by GPS for land vehicle applications. *J. Navigation*, 63:663–680, 2010.

100. Y. Hong, G. Chen, and L. Bushnell. Distributed observers design for leader-following control of multi-agent networks. *Automatica*, 44(3):846–850, 2008.

101. Y. Hong, J. Hu, and L. Gao. Tracking control for multi-agent consensus with an active leader and variable topology. *Automatica*, 42(7):1177–1182, 2006.

102. J. Hu and Y. Hong. Leader-following coordination of multi-agent systems with coupling time delays. *Physica A*, 374:853–863, 2007.

103. L. Hu, P. Shi, and P. M. Frank. Robust sampled-data control for markovian jump linear systems. *Automatica*, 42(11):2025–2030, 2006.

104. Q. L. Hu. Variable structure maneuvering control with time-varying sliding surface and active vibration damping of flexible spacecraft with input saturation. *Acta Astronaut.*, 64:1085–1108, 2009.

105. Q. L. Hu and G. F. Ma. Variable structure control and active vibration suppression of flexible spacecraft during attitude maneuver. *Aerosp. Sci. Technol.*, 9:307–317, 2005.

106. J. Huang and Z. Chen. A general framework for tackling the output regulation problem. *IEEE Trans. Automat. Contr.*, 49(12):2203–2218, 2004.

107. L. Huang. *The Theoretical Basis of the Stability and Robustness (in Chinese)*. Science Press, Beijing, 2002.

108. Y. H. Huang and W. Massner. A novel disturbance observer design for magnetic hard drive servo system with rotary actuator. *IEEE Trans. Magn.*, 34:1892–1894, 1998.

109. R. Isermann and P. Balle. Trends in the application of model based fault detection and diagnosis of technical process. pages 1–12, San Francisco, CA, 1996. In

Proc. 13th World Congr. Int. Federation Automatic Control (IFAC'96).

110. J. Ishikawa and M. Tomizuka. Pivot friction compensation using an accelerometer and a disturbance observer for hard disk. *IEEE/ASME Trans. Mech.*, 3:194–201, 1998.

111. A. Isidori. *Nonlinear Control Systems.* Springer, Berlin, 1995.

112. A. Isidori and C. I. Byrnes. Output regulation of nonlinear systems. *IEEE Trans. Automat. Contr.*, 35(2):131–140, 1990.

113. M. Iwasaki, T. Shibata, and N. Matsui. Disturbance-observer-based nonlinear friction compensation in the table drive systems. *IEEE/ASME Trans. Mech.*, 4(1):3–8, 1999.

114. A. Jadbabaie, J. Lin, and A. S. Morse. Coordination of groups of mobile agents using nearest neighbor rules. *IEEE Trans. Automat. Contr.*, 48(6):988–1001, 2003.

115. B. Jiang and F. N. Chowdhury. Fault estimation and accommodation for linear MIMO discrete time systems. *IEEE Trans. Contr. Syst. tech.*, 13:493–499, 2005.

116. B. Jiang, M. Staroswiecki, and V. Cocquempot. Active fault tolerant control for a class of nonlinear systems. pages 127–132, Washington, 2003. Proc. IFAC Safeprocess.

117. B. Jiang, K. Zhang, and P. Shi. Less conservative criteria for fault accommodation of time-varying delay systems using adaptive fault diagnosis observer. *Int J. Adapt. Control*, 24:322–334, 2010.

118. Y. F. Jiang. Error analysis of analytic coarse alignment methods. *IEEE Trans. Aero. Elec. Sys.*, 34(1):334–337, 1998.

119. X. Z. Jin and G. H. Yang. Robust fault-tolerant control systems design with actuator failures via linear fractional transformations. pages 4338–4343, Seattle, WA, 2008. Proc. 2008 Amer. Contr. Conf.

120. D. Karagiannis, M. Sassano, and A. Astolfi. Dynamic scaling and observer design with application to adaptive control. *Automatica*, 45(9):2883–2889, 2009.

121. C. J. Kemf and S. Kobayashi. Disturbance observer and feedforward design for a highspeed directdrive positioning table. *IEEE Trans. Contr. Syst. tech.*, 7(5):513–526, 1999.

122. B. K. Kim and W. K. Chung. Unified analysis and design of robust disturbance attenuation algorithms using inherent structural equivalence. volume 25, pages 4046–4051, Arlington, VA, 2001. Proc. of the ACC.

123. B. K. Kim, W. K. Chung, and Y. Youm. Robust learning control for robot manipulators based on disturbance observer. pages 1276–1282. In Proc. IEEE IEC, 1996.

124. E. Kim and S. Lee. Output feedback tracking control of MIMO systems using a fuzzy disturbance observer and its application to the speed control of a PM synchronous motor. *IEEE Trans. Fuzzy Syst.*, 13(6):725–741, 2005.

125. S. Komada and K. Ohnishi. Force feedback control of robot manipulator by the acceleration tracing orientation method. *IEEE Trans. Ind. Electron.*, 37:6–12, 1990.

126. C. Kravarisa and V. Sotiropoulosa. Nonlinear observer design for state and

disturbance estimation. *Syst. Control Lett.*, 56:730–735, 2007.

127. M. Krstic and H. Deng. *Stabilisation of Nonlinear Uncertain Systems*. Springer, Berlin, 1998.

128. J. Lam, Z. Shu, S. Xu, and E. K. Boukas. Robust H_∞ control of descriptor discrete-time markovian jump systems. *Int. J. Control*, 80(3):374–385, 2007.

129. D. Lee and M. W. Spong. Agreement with non-uniform information delays. pages 756–761, Minneapolis, MN, 2006. Proc. 2006 Amer. Contr. Conf.

130. D. Lee and M. W. Spong. Stable flocking of multiple inertial agents on balanced graphs. *IEEE Trans. Automat. Contr.*, 52(8):1469–1475, 2007.

131. H. K. Lee and J. G. Lee. Calibration of measurement delay in global positioning system/strapdown inertial navigation system. *J. Guid. Control Dyn.*, 25(2):240–247, 2002.

132. F. L. Lewis and D. M. Dawson. *Robot Manipulator Control: Theory and Practice*. Marcel Dekker, Inc., New York, 2004.

133. H. X. Li and P. P. J. Van Den Bosch. A robust disturbance-based control and its application. *Int. J. Control*, 58:537–554, 1993.

134. T. Li and L. Guo. Optimal fault-detection filtering for non-gaussian systems via output PDFs. *IEEE Trans. Syst. Man Cy.-A*, 39(2):476–481, 2009.

135. X. Li, X. F. Wang, and G. R. Chen. Pinning a complex dynamical networks to its equilibrium. *IEEE Trans. Circuits-I*, 51(10):2074–2087, 2004.

136. C. L. Lin and T. Y. Lin. An H_∞ design approach for neural net-based control schemes. *IEEE Trans. Automat. Contr.*, 46:1599–1605, 2001.

137. C. S. Liu and H. Peng. Disturbance observer based tracking control. *J. Dyn. Syst-T ASME*, 122:332–335, 2000.

138. Z. L. Liu and J. Svoboda. A new control scheme for nonlinear systems with disturbances. *IEEE Trans. Contr. Syst. Technol.*, 14(1):176–181, 2006.

139. S. L. Lü, L. Xie, and J. B. Chen. New techniques for initial alignment of strapdown inertial navigation system. *J. Franklin Inst.*, 346:1021–1037, 2009.

140. M. S. Mahmoud, Y. Shi, and H. N. Nounou. Resilient observer-based control of uncertain time-delay systems. *Int. J. of Innov. Comput. Inform.*, 3(2):407–418, 2007.

141. Z. H. Man and X. H. Yu. Terminal sliding mode control of MIMO linear systems. *IEEE Tran. Circuits-I*, 44:1065–1070, 1997.

142. R. Marino and G. L. Santosuosso. Global compensation of unknown sinusoidal disturbances for a class of nonlinear nonminimum phase systems. *IEEE Trans. Automat. Contr.*, 50(11):1816–1822, 2005.

143. R. Marino, G. L. Santosuosso, and P. Tomei. Robust adaptive observers for nonlinear systems with bounded disturbances. *IEEE Trans. Automat. Contr.*, 46(6):967–972, 2001.

144. R. Marino and P. Tomei. *Nonlinear Control Design: Geometric, Adaptive and Robust*. Prentice-Hall, Englewood Cliffs, NJ, 1995.

145. R. Marino and P. Tomei. Output regulation for linear systems via adaptive internal model. *IEEE Trans. Automat. Contr.*, 48(12):2199–2202, 2003.

146. R. Marino and P. Tomei. Adaptive tracking and disturbance rejection for uncer-

tain nonlinear systems. *IEEE Trans. Automat. Contr.*, 50(1):90–95, 2005.

147. F. Maurizio and B. A. Foued. Adaptive regulation of MIMO linear systems against unknown sinusoidal exogenous inputs. *Int. J. Adapt. Control*, 23(6):581–603, 2009.

148. D. McRuer, I. Ashkenas, and D. Graham. *Aircraft Dynamics and Automatic Control*. Princeton University Press, Princeton, 1976.

149. A. J. Miller, G. L. Gray, and A. P. Mazzoleni. Nonlinear spacecraft dynamics with flexible appendage, damping, and moving internal submasses. *J. Guid. Control Dyn.*, 42(3):605–615, 2001.

150. T. Mita, M. Hirata, and K. Murata. H_∞ control versus disturbance-observer-based control. *IEEE Trans. Ind. Electron.*, 45:488–495, 1998.

151. L. Moreau. Stability of continuous-time distributed consensus algorithms. volume 4, pages 3998–4003, Nassau, Bahamas, 2004. in Proc. 43rd IEEE CDC.

152. L. Moreau. Stability of multi-agent systems with time dependent communication links. *IEEE Trans. Automat. Contr.*, 50(2):169–182, 2005.

153. P. C. Muller and J. Ackermann. Nichtlineare regelung von elastischen robotern. *VDI-Berichte*, 598:321–333, 1986.

154. T. Nagata, V. J. M. Odi, and H. Matsuo. Dynamics and control of flexible multibody systems part I: general formulation with an order N forward dynamics. *Acta Astronaut.*, 49(11):581–594, 2001.

155. M. Nakao, K. Ohnishi, and K. Miyachi. A robust decentralized joint control based on interference estimation. pages 326–331. In Proc. IEEE Int. Conf. Robotics and Automation, 1987.

156. E. Nebot and H. Durrant-Whyte. Initial calibration and alignment of low cost inertial navigation units for land vehicle applications. *J. Robotics Systems*, 16(2):81–92, 1999.

157. H. Nijmeijer and A. J. Van Der Schaft. *Nonlinear Dynamical Control Systems*. Springer, Berlin, 3rd edition, 1996.

158. V. Nikiforov. Adaptive nonlinear tracking with complete compensation of unknown disturbances. *Eur. J. Control*, 4(2):132–139, 1998.

159. V. Nikiforov. Nonlinear servo compensation of unknown external disturbances. *Automatica*, 37(10):1647–1653, 2001.

160. Y. Niu, D. W. C. Ho, and X. Wang. Sliding mode control for Itô stochastic systems with markovian switching. *Automatica*, 43(10):1784–1790, 2007.

161. N. Oda, T. Murakami, and K. Ohnishi. A robust control strategy of redundant manipulator by workspace observer. *J. Robot. Mechatron.*, 8:235–242, 1996.

162. Y. Oh and W. K. Chung. Disturbance-observer-based motion control of redundant manipulators using inertially decoupled dynamics. *IEEE/ASME Trans. Mechatron.*, 4:133–145, 1999.

163. R. Olfati-Saber, J. A. Fax, and R. M. Murray. Consensus and cooperation in networked multi-agent systems. volume 95, pages 215–233. Proc. of the IEEE, 2007.

164. R. Olfati-Saber and R. M. Murray. Consensus problems in networks of agents with switching topology and time-delays. *IEEE Trans. Automat. Contr.*,

49(9):1520–1533, 2004.

165. M. C. De Oliveira, J. Bernussoub, and J. C. Geromela. A new discrete-time robust stability condition. *Syst. Control Lett.*, 37:261–265, 1999.

166. S. Oucheriah. Global stabilization of a class of linear continuous time-delay systems with saturating controls. *IEEE Trans. Circuits-I*, 43:1012–1015, 1996.

167. I. R. Petersen and C. V. Hollot. A riccati equation approach to the stabilization of unceratin linear systemss. *Automatica*, 22:397–411, 1986.

168. M. M. Polycarpou and A. B. Trunor. Learning approach to nonlinear fault diagnosis: detectability analysis. *IEEE Trans. Automat. Contr.*, 45:806–812, 2000.

169. W. J. Qing. Disturbance rejection through disturbance observer with adaptive frequency estimation. *IEEE Trans. Magn.*, 45(6):2675–2678, 2009.

170. A. Radke and Z. Gao. A survey of state and disturbance observers for practitioners. pages 14–16, Minneapolis, MN, 2006. Proc. 2006 Amer. Contr. Conf.

171. W. Ren. Consensus strategies for cooperative control of vehicle formations. *IET Control Theory Appl.*, 1(2):505–512, 2007.

172. W. Ren and E. Atkins. Distributed multi-vehicle coordinated control via local information exchange. *Int. J. Robust Nonlin.*, 17(11):1002–1033, 2007.

173. W. Ren and R. W. Beard. Consensus seeking in multi-agent systems under dynamically changing interaction topologies. *IEEE Trans. Automat. Contr.*, 50(9):655–661, 2005.

174. W. Ren and R. W. Beard. *Distributed Consensus in Multi-Vehicle Cooperative Control: Theory and Applications*. Springer-Verlag, London, 2008.

175. W. Ren, R. W. Beard, and E. Atlkins. Information consensus in multivehicle cooperative control: collective group behavior through local interaction. *IEEE Contr. Syst. Mag.*, 27(2):71–82, 2007.

176. W. Ren and Y. C. Cao. *Distributed Coordination of Multi-agent Networks*. Springer-Verlag.

177. R. M. Rogers. *Applied Mathematics in Integrated Navigation Systems*. AIAA Inc., Virginia, USA, 3rd edition, 2007.

178. B. Scherzinger. Inertial navigator error models for large heading uncertainty. pages 477–484, Atlanta, USA, 1996. Proc. IEEE Position Location and Navigation Symposium.

179. E. Schrijver and V. J. Dijk. Disturbance observers for rigid mechanical systems: Equivalence, stability, and design. *J. Dyn. Syst-T ASME*, 124:539–548, 2002.

180. A. Serrani. Rejection of harmonic disturbances at the controller input via hybrid adaptive external models. *Automatica*, 42:1977–1985, 2006.

181. A. Serrani, A. Isidori., and L. Marconi. Semiglobal nonlinear output regulation with adaptive internal model. *IEEE Trans. Automat. Contr.*, 46(8):1178–1194, 2001.

182. J. She, Y. Ohyama, and M. Nakano. A new approach to the estimation and rejection of disturbances in servo systems. *IEEE Trans. Contr. Syst. Technol.*, 13(3):378–385, 2005.

183. T. L. Shen. H_∞ *Control Theory and Application (in Chinese)*. Tsinghua Uni-

versity Press, Beijing, 1996.

184. P. Shi, Y. Xia, G. Liu, and D. Rees. On designing of sliding-mode control for stochastic jump systems. *IEEE Trans. Automat. Contr.*, 51(1):97–103, 2006.

185. M. J. Sidi. *Spacecraft Dynamics and Control*. Cambridge University Press, UK, 2000.

186. J. E. Slotine and W. Li. *Applied Nonlinear Control*. Prentice-Hall, Englewood Cliffs, NJ, 1991.

187. G. Song and B. Kotejoshyer. Vibration reduction of flexible structures during slew operations. *Int. J. Acoust. Vib.*, 7(2):105–109, 2002.

188. Y. X. Su, C. H. Zheng, and B. Y. Duan. Automatic disturbances rejection controller for precise motion control of permanent-magnet synchronous motors. *IEEE Trans. Ind. Electron.*, 52:814–823, 2005.

189. M. Sun, J. Lam, S. Xu, and Y. Zou. Robust exponential stabilization for markovian jump systems with mode-dependent input delay. *Automatica*, 43(10):1799–1807, 2007.

190. X. M. Sun, J. Zhao, and W. Wang. Two design schemes for robust adaptive control of a class of linear uncertain neutral delay systems. *Int. J. Innov. Comput. Inform.*, 3(2):385–396, 2007.

191. Y. Sun, L. Wang, and G. Xie. Average consensus in networks of dynamic agents with switching topologies and multiple time-varying delays. *Syst. Control Lett.*, 57(2):175–183, 2008.

192. Y. Tang. Terminal sliding mode control for rigid robots. *Automatica*, 34:51–56, 1998.

193. M. Tomizuka and K. K. Chew. Disturbance rejection through an external model. *J. Dyn. Syst-T ASME*, 112(4):559–564, 1990.

194. S. C. Tu. *Satellite Attitude Dynamics and Control (in Chinese)*. China Astronatica Publishing House, Beijing, 2001.

195. T. Umeno and Y. Hori. Robust speed control of DC servomotors using modem two degree of freedom controller design. *IEEE Trans. Ind. Electron.*, 38:363–368, 1991.

196. V. I. Utkin. *Sliding Modes in Control Optimization*. Springer, Berlin, 1992.

197. C. C. Wang and M. Tomizuka. Design of robustly stable disturbance observers based on closed loop consideration using H_∞ optimization and its applications to motion control systems. pages 3764–3769, Boston, MA, 2004. Proc. 2004 Amer. Contr. Conf.

198. L. X. Wang and L. Guo. Multi-objective control in integrated navigation system based on LMI. *Acta Aeronau. Astronau. Sinica*, 29S:102–106, 2008.

199. X. L. Wang. Fast alignment and calibration algorithms for inertial navigation system. *Aerosp. Sci. Technol.*, 13:204–209, 2009.

200. Z. Wang, J. Lam, and X. Liu. Nonlinear filtering for state delayed systems with markovian switching. *IEEE Trans. Signal Process.*, 51(9):2321–2328, 2003.

201. X. J. Wei and L. Guo. Composite disturbance-observer-based control and terminal sliding mode control for non-linear systems with disturbances. *Int. J. Control*, 82(6):1082–1098, 2009.

202. X. J. Wei and L. Guo. Composite disturbance-observer-based control and H_∞ control for complex continuous models. *Int. J. Robust Nonlin.*, 20:106–118, 2010.

203. X. J. Wei, H. F. Zhang, and L. Guo. Composite disturbance-observer-based control and variable structure control for non-linear systems with disturbances. *Trans. Inst. Meas. Control*, 31(5):401–423, 2009.

204. X. Y. Wen and L. Guo. Attenuation and rejection for multiple disturbances of nonlinear robotic systems using nonlinear observer and PID controller. pages 2512–2517, Jinan, China, 2010. Proc. of the 8th WCICA.

205. H. Wu. Continuous adaptive robust controllers guaranteeing uniform ultimate boundness for uncertain nonlinear systems. *Int. J. Control*, 72:115–122, 1999.

206. H. X. Wu, J. Hu, and Y. C. Xie. *Intelligent Adaptive Control Based on Characteristics Model (in Chinese)*. Chinese Science and Technology Press, Beijing, 2009.

207. Y. Q. Wu, X. H. Yu, and Z. H. Man. Terminal sliding mode control design for uncertain dynamic systems. *Syst. Control Lett.*, 34:281–287, 1998.

208. Z. R. Xi and Z. Ding. Global adaptive output regulation of a class of nonlinear systems with nonlinear exosystems. *Automatica*, 43(1):143–149, 2007.

209. Y. Xia, J. Zhang, and E. K. Boukas. Control for discrete singular hybrid systems. *Automatica*, 44(10):2635–2641, 2008.

210. B. Xian and N. Jalili. Adaptive tracking control of linear uncertain mechanical systems subjected to unknown sinusoidal disturbances. *J. Dyn. Syst-T ASME*, 125(3):129–134, 2003.

211. F. Xiao and L. Wang. Asynchronous consensus in continuous-time multi-agent systems with switching topology and time-varying delays. *IEEE Trans. Automat. Contr.*, 53(8):1804–1816, 2008.

212. J. Xiong and J. Lam. Fixed-order robust H_∞ filter design for markovian jump systems with uncertain switching probabilities. *IEEE Trans. Signal Process.*, 54(4):1421–1430, 2006.

213. J. Xiong and J. Lam. Stabilization of discrete-time markovian jump linear systems via time-delayed controllers. *Automatica*, 42(5):747–753, 2006.

214. F. Xu and J. C. Fang. Velocity and position error compensation using strapdown inertial navigation system/celestial navigation system integration based on ensemble neural network. *Aerosp. Sci. Technol.*, 12:302–307, 2008.

215. J. M. Xu, Q. J. Zhou, and T. P. Leung. Implicit adaptive inverse control of robot manipulators. pages 334–339, Atlanta, 1993. Proc. of the IEEE Conf. Robotics and Automation.

216. F. W. Yang, Z. D. Wang, and M. Gani. Robust H_2/H_∞ filtering for uncertain systems with missing measurement. pages 4769–4774, San Diego, USA, 2006. Proc. of the 45th IEEE CDC.

217. H. Yang and S. Zhang. Consensus of multi-agent moving systems with heterogeneous communication delays. *Int. J. Syst. Contr. Commun.*, 2(1):426–436, 2009.

218. H. Y. Yang, S. Zhang, and G. D. Zong. Trajectory control of scale-free dynamical

networks with exogenous disturbance. *Commun. Theor. Phys.*, 55(1):185–192, 2011.

219. J. Yang and S. H. Li. Disturbance rejection of ball mill grinding circuits using DOB and MPC. *Powder Technol.*, 198:219–228, 2010.

220. P. Yang, R. A. Freeman, and K. M. Lynch. Multi-agent coordination by decentralized estimation and control. *IEEE Trans. Automat. Contr.*, 53(11):2480–2496, 2008.

221. Z. Yang and H. Tsubakihara. A novel robust nonlinear motion controller with disturbance observer. *IEEE Trans. Contr. Syst. Technol.*, 16(1):137–147, 2008.

222. X. Yao, L. Wu, and W. X. Zheng. Fault detection for discrete-time markovian jump singular systems with intermittent measurements. *IEEE Trans. Signal Process.*, 59(7):3099–3109, 2011.

223. X. Yao, L. Wu, W. X. Zheng, and C. Wang. Robust H_∞ filtering of markovian jump stochastic systems with uncertain transition probabilities. *Int. J. Syst. Sci.*, 42(7):1219–1230, 2011.

224. D. Ye and G. H. Yang. Adaptive fault-tolerant tracking control against actuator faults with application to flight control. *IEEE Trans. Contr. Syst. Technol.*, 14(6):1088–1096, 2006.

225. L. Yu, G. D. Chen, and M. R. Yang. Robust reonal pole assignment of uncertain systems via output feedback controller. *Contr. Theor. Appl.*, 9(2):244–246, 2002.

226. M. J. Yu, J. G. Lee, and C. G. Park. Nonlinear robust observer design for strapdown INS in-flight alignment. *IEEE Trans. Aero. Elec. Sys.*, 40(3):797–807, 2004.

227. M. J. Yu and S. W. Lee. A robust extended filter design for SDINS in-flight alignment. *Int. J. Control, Autom.*, 1(4):520–526, 2003.

228. W. Yu, G. R. Chen, and J. Lv. On pinning synchronization of complex dynamical networks. *Automatica*, 45:429–435, 2009.

229. Q. H. Zhang, M. Basseville, and A. Benveniste. Fault detection and isolation in nonlinear dynamic systems: a combined input-output and local approach. *Automatica*, 34:1359–1373, 1998.

230. Y. F. Zhang and Y. Gao. A method to improve the alignment performance for GPS-IMU system. *GPS Solut.*, 11:129–137, 2007.

231. Y. M. Zhang, L. Guo, and H. Wang. Robust filtering for fault tolerant control using output PDFs of non-Gaussian systems. *IET Control Theory Appl.*, 1:636–645, 2007.

232. Y. M. Zhang and J. Jiang. Bibliographical review on reconfigurable fault-tolerant control systems. *Annu. Rev. Control*, 32:229–252, 2008.

233. Z. H. Zhang and Z. D. Xu. Time-variant linear system of micro hard disk driver and its track-seeking. pages 6288–6292, Dalian, China, 2006. Proc. of the 6th WCICA.

234. Y. Zheng, C. Zhang, and R. J. Evans. A differential vector space approach to nonlinear system regulation. *IEEE Trans. Automat. Contr.*, 45:1997–2010, 2000.

235. M. Zhong, H. Ye, P. Shi, and G. Wang. Fault detection for markovian jump

systems. *IEE P. Contr. Theor. Ap.*, 152(4):397–402, 2005.

236. M. Y. Zhong, Y. X. Liu, and Z. Y. Huo. H_∞ fault estimation and accommodation for LTI systems. pages 1806–1809, Chongqing, China, 2008. Proc. 7th World Conf. Intelligent Control and Automation.

237. Q. C. Zhong and J. E. Normey-Rico. Disturbance observer-based control for processes with an integrator and long dead-time. volume 3, pages 2261–2266. Proc. of the 40th IEEE CDC, 2001.

238. K. M. Zhou and J. C. Doyle. *Essentials of Robust Control*. Prentice Hall, New Jersey, 1998.

239. X. M. Zhu, C. C. Hua, and S. Wang. State feedback controller design of networked control systems with time delay in the plant. *Int. J. of Innov. Comput., Inform.*, 4(2):283–290, 2008.

Index